Beyond Guns and Steel

Beyond Guns and Steel

A War Termination Strategy

DOMINIC J. CARACCILO

Foreword by Daniel P. Bolger

An AUSA Book

Praeger Security International

 PRAEGER

AN IMPRINT OF ABC-CLIO, LLC
Santa Barbara, California • Denver, Colorado • Oxford, England

Library of Congress Cataloging-in-Publication Data

Caraccilo, Dominic J. (Dominic Joseph), 1962-
  Beyond guns and steel: a war termination strategy / Dominic J. Caraccilo; foreword
  by Daniel P. Bolger.
      p. cm.
"An AUSA Book."
  Includes bibliographical references and index.
    ISBN 978-0-313-39149-1 (hard copy: alk. paper)—ISBN 978-0-313-39150-7 (ebook)
  1. Military planning—United States. 2. War—Termination. 3. Disengagement
(Military science) I. Title.
    U153.C368   2011
    355.02′8—dc22   2010040757

ISBN: 978-0-313-39149-1
EISBN: 978-0-313-39150-7

15   14   13   12   11   1   2   3   4   5

This book is also available on the World Wide Web as an eBook.
Visit www.abc-clio.com for details.

Praeger
An Imprint of ABC-CLIO, LLC

ABC-CLIO, LLC
130 Cremona Drive, P.O. Box 1911
Santa Barbara, California 93116-1911

This book is printed on acid-free paper ∞

Manufactured in the United States of America

For the One

Contents

Foreword

Are we there yet?

What parent has not heard that line wander up from the back seat, or thought it for himself or herself, while droning ahead on a long, lonesome highway? An automobile journey has a destination, and, sooner or later, you reach the end. Therefore, when the question arises, the aware driver can answer with an estimated arrival time. But all of that presupposes that you know where you are going.

One would like to think that when a country goes to war, it knows where it wants to end up. In contemporary American strategic thought, much time and energy is devoted to putative "endstates" and victory conditions. But, not unlike our generic car trip, the route chosen must get you to the final stop. One cannot drive to Las Vegas by going north out of New York City, and we would scoff mightily at a vehicle operator who tried such a maneuver. It works only when you pick an objective that connects to your approach, and it works even better if the path selected is smooth, straight, and short.

Yet the United States, like other countries in history, has at times moved well down the road before determining that something has gone seriously wrong with the strategic compass . . . not to mention the moral one. In certain circumstances, we have marched off to fight without fully thinking through where we really should go and how best we could get there. One thinks of the War of 1812, the Philippine Insurrection, the Korean War, the Vietnam War, and the current campaigning in the wake of the terror attacks of September 11, 2001. To have any hope of justifying the blood sacrifices it demands, a war must end well.

This is exactly the subject of this book, and a timely one it is. Author Dominic Caraccilo has "seen the elephant," as Johnny Reb and Billy Yank put it during the Late Unpleasantness of 1861 to 1865. Even in an army with a lot of combat experience, he has accumulated more than most, with recent service in both Afghanistan and Iraq. Moreover, he has thought about what he has experienced and written about it. He's a thinking soldier. Lately, he has been thinking about the most important thing of all: how to end a war.

The author's ideas are grounded in the realities of today's conflict against Al Qaeda terrorists and their ilk. This is the war he has fought in, researched, and assessed. And this is the war that he explains how to end.

You can end a war at any point, just as you can end a car trip at any point. Just stop and get out. Or worse, turn around and go home. Or worst of all, spin the wheel and head for the ditch. But those are not outcomes worth a book. This account is not about quitting; it is about finishing a very hard journey and making it worth the trip.

Are we there yet? Not yet. But Dom Caraccilo has shown us the way.

Daniel P. Bolger
Lieutenant General
G3 (Operations Officer of the U.S. Army)

Introduction

There are two things which will always be difficult for a democratic nation to do: beginning and ending a war.

—Alexis de Tocqueville, *Democracy in America*

The summer of 2009 reeks of heat and stagnation in Baghdad, Iraq. Technically, things are good: attacks on Coalition Forces have nearly bottomed out; the Iraqi army and police conduct autonomous operations and display professionalism to an extent that seemed impossible only a few years ago. The Iraqis have signed a security agreement with the United States that puts them in the lead and moves the U.S. troops out of the cities. The commander of forces, General Raymond T. Odierno, has devised a plan, in concert with President Barack Obama's order, to reduce combat forces to 50,000 by September 2010 and end "combat operations" and eventually withdraw all of the troops by the end of 2011.

However, like the scorching winds that sweep in from the desert and push across the fertile land between the rivers, success has brought more frustration than relief. The war machine keeps rolling, but with an enemy that has mostly dissipated and foreign internal defense taking many matters into Iraqi hands, ennui settles heavily on the hundreds of patrol bases crouching in the desert. Purpose evades the ever-moving soldiers the way the shadowy insurgent enemy once did. Victory seems, in so many ways, achieved, or at the very least, achievable, so what is left to do in a war whose main objectives have been accomplished? There is no doubt that with the reduction of combat forces to support the responsible drawdown of forces that there is much for the individual soldier to do. Regardless, for the U.S. soldiers serving in Operation Iraqi Freedom (OIF), this place beyond victory shifts amoeba-like from hour to hour.[1]

This is how the war in Iraq felt in the summer of 2009 and at other salient moments since it began in 2003. The frustration of the soldiers, many of whom had multiple tours and saw the progression or lack thereof, was that there was never a plan in place to achieve victory and, regardless of the current progress, a sense of doubt was prevalent. Like other places in Iraq, six different combat formations in the past seven years had occupied the same ground for at least a 12-month interval. It leaves one questioning, what was the plan? Did the architects of this war have an endstate in mind that included more than just the defeat of enemy factions? Did the plan include a reconstruction effort for the country it had first sanctioned for years and then invaded? What the reader may find shocking is that although the U.S. security apparatus has an exceedingly well-developed plan to fight conventional wars and an inherent ability to fight unconventional wars, as most recently shown in the Petraeus-led writing of the 2007 *Field Manual 3-24: Counterinsurgency Operations* and the 2007 Odierno-led surge in Iraq, it lacks a plan for what is doctrinally known as *war termination* and *conflict termination*.

There is little doubt that it is always easier to get into a fight than to get out of one. Four times in the last century the United States has come to the end of a major war. Each time it concluded that the nature of man and the world had changed for the better, turned inward unilaterally, and dismantled institutions important to our national security, giving ourselves a so-called "peace" dividend. Four times we chose to forget history.[2] Instead of developing a war termination or exit strategy, or even more specially, a postwar reconstruction strategy, we went on to the next conflict ignoring the results of the previous foray. More importantly, we ignored how we improperly planned for the previous war to end.

Does war ever just end? Is there a so-called endstate? Or does it evolve? And if it evolves, does it do so naturally or does leadership, or perhaps an exhaustive plan, make the transition from war to peace simpler and more efficient? What happens if the leader is weak and the plan is lacking? What happens to the indigenous population affected by the war? If there is no plan and the leaders are weak, is the natural outcome for the ill-fated inhabitants a political and domestic crisis? If so, then shouldn't that result be exactly what we should plan on preventing even before the guns begin to shoot? This book will make an effort to answer these questions and more. It is a look at what happens beyond the guns and steel, beyond bullets and sweat, beyond the confusion and angst of war. My hope is that it makes those in public service think before committing our nation's military might and that those in military service respect the challenges brought on by war.

Is it true that the injection of the right leader at the right time in any conflict can remedy any flailing war footing? The introduction of General Ray Odierno as the Multi-National Corps–Iraq (MNC-I) commander in 2007 is an example of "righting the ship" when he introduced a population-centric counterinsurgency tactic replete with a complex but effective reconciliation effort to turn the tide in what was certainly a brutal civil war motivated by violence brought on by sectarianism and terrorism. Even without a plan ahead of time, there is no doubt success can be achieved, but it will be at a cost. The goal is to diminish the cost.

Known inside think-tank circles as "The Patton of Counterinsurgency," General Ray Odierno took command of MNC-I (and as of this writing the commander of U.S. Force–Iraq or USF-I) on December 14, 2006. Iraq was in flames. Insurgents and death squads were killing 3,000 civilians a month. Coalition Forces were sustaining more than 1,200 attacks per week. Operation Together Forward II, the 2006 campaign to clear Baghdad's most violent neighborhoods and hold them with Iraqi Security Forces, had been suspended because violence elsewhere in the capital was rising steeply. Al Qaeda in Iraq owned safe havens within and around Baghdad, throughout Anbar, and in Diyala, Salah-ad-Din, and Ninewa provinces. The Iraqi government was completely paralyzed, and there was no end in sight.

When General Odierno relinquished command of MNC-I on February 14, 2008, the civil war was over. Civilian casualties were down 60 percent as were weekly attacks. By June 30, 2009, when General Odierno, who was a now a four star general and back in Iraq as commander of all forces in Iraq (MNC-I and on January 1, 2010, USF-I), the security situation was so stable that U.S. Forces could turn over the cities to the Iraqi Security Forces; by September 2010, the U.S. Forces was, in turn, reduced to a mere 50,000—a fraction of what had been on the ground less than two years prior.[3] This is an example of how one man can make a difference and how leadership, the most dynamic element of combat power, can change the course of battle. As the great Prussian military analyst Carl von Clausewitz, recognized. "Even the ultimate outcome of war is not always to be regarded as final. The defeated state often considers the outcome merely as a transitory evil, for which a remedy may still be found in political conditions at some later date."[4] However, this book is about how to alleviate the need for "the knight in shiny armor coming to the rescue." Instead we want to plan for it ahead of time so that rescuing is not needed.

EXAMINING WAR TERMINATION AS PART OF STRATEGY[5]

Two things routinely happen when a nation decides to employ its full array of power into combat. Those engaged in conflict can either (1) follow a predeveloped, completely resourced, and fully synchronized plan for reconstruction and stability or nation-building as an integral part of combat operations and not necessarily wait until the fight comes to an end to define what to do when the fighting stops; or (2) if the war termination and conflict resolution plan is weak, as it usually is, or the resources are lacking, and they usually are, then those engaged in combat will have to define how to evolve or terminate the war.

The next logical step is a transition to peace or, at least, stability, all the while competing for resources at a time when the interest in the "conflict" has habitually waned by the public.[6] Lest we not forget, it is the public (or the constituents) who hold the purse strings in Congress. One does not have to look further back than 1993 to see how the public outrage for the 18 American loses in Somalia caused a president to immediately retreat (see Chapter 4).

Having a well-developed and exhaustive plan is better by an order of magnitude than coming up with one on the fly in the nth hour. Unfortunately, more times than not, history repeats itself (or, as Mark Twain completes the sentence, ". . . or at least rhymes"), and those soldiers engaged in all phases of the fight are left with the responsibility to figure out what comes next. Although the leaders in today's military have the experience and capacity to creatively pull off a success, that hasn't always been the case in history. I am less inclined to believe the same positive potential will exist in the future.

Strategy expert and author George Friedman writes in *The Next 100 Years: A Forecast for the 21st Century* that, "In geopolitics, major conflicts repeat themselves. France and Germany, for example, fought multiple wars, as did Poland and Russia. . . . Significant conflicts are rooted in underlying realities. . . . Keep in mind how quickly Balkan geopolitics lead to a recurrence of wars that been fought a century earlier."[7] If one follows this line of thinking, what is more troublesome is that it seems that history tends to repeat itself in the forming of conflict, but has absolute amnesia when devising a plan for terminating war and conflict resolution. The exception may be thematic as seen in the "Balkanization" effort in OIF, where one-for-one unit replacements were placed against stalled mission-sets.[8] However, the Balkanization effort is more about how security forces are sustained then bringing about the end of war and its post-conflict resolution.

It is my intent in this work to provide an artful examination of military strategies existing in a nation's arsenal for planning and then executing the outcome of war. The lens I use to frame and then view my argument is my personal experiences in more than six years of combat deployments—first as a company commander in Iraq in Desert Storm with the 82nd Airborne Division; as a U.S. Army Ranger in Afghanistan at the onset of Operation Enduring Freedom (OEF); as an airborne battalion commander in Kosovo during Operation Rapid Guardian, and then a month later deploying via parachute into Kurdistan for the beginning of OIF; as that same battalion commander living in the ethnically explosive city of Kirkuk for a year; as the Division Operations officer for the 101st Airborne Division, or Multi-Division North in Northern Iraq at the height of the insurgency for another year; as a Brigade Combat Team (BCT) Commander in South Baghdad in the heart of the former "Triangle of Death" for 15 months; and then finally, as the Executive Officer to the Multi-National Force Iraq (and then the transitioned USF-I) Commander as the war in Iraq, or at least the U.S. military participation, came to a close.

I will show later that when discussing exit strategies I am defining an era's *grand strategy* for conflict. Although the potential for numerous arguments exist, this book's main focus is exploring this nation's ability, or in most cases, inability to set clear and concise objectives as it embarks on a fight and employs its awesome power. *Beyond Guns and Steel* describes power not only in military terms, but as power that includes a full array of interagency involvement from diplomacy and economic and information infusion. Among Foreign Service Officers (FSOs) and military leaders, this power is known as DIME (diplomacy, information, military, economic). So the construct exists, at least in the definitions. The D, I, and E portions of DIME are widely known as *soft* or *smart* power. The M, or military, is often seen as the *hard* power.

Much has been written on the contrasting definitions of *power*. Hendrick Hertzberg in the April 20, 2009, edition of *The New Yorker* describes the difference of hard power—military force, soft power—essentially diplomacy, and the synthesis of both in a new wave definition as *smart power*. This kind of power is favored by Secretary of State Hillary Clinton in her opening statement at her confirmation hearing before the Senate Foreign Relations Committee:

> We must use what has been called smart power, the full range of tools at our disposal—diplomatic, economic, military, political, legal, and cultural—picking the right tool, or combination of tools, for each situation. With smart power, diplomacy will be the vanguard of our foreign policy. This is not a radical idea. The ancient Roman poet Terence declared that "in every endeavor, the seemly course for wise men is to try persuasion first . . ."[9]

When deciding on what type of power to employ in conflict; one must always remember that soldiers employ combat power and diplomats persuade. Perhaps we are at a point where smart, hard, and soft power are spun together in a web of resources and labeled *national power.* Clearly the linchpin to success if this were the case is the congressional backing for resources—a struggle that limits all areas of power except the defense departments stranglehold on the budgeting for hard power.

Beyond defining the types of power a nation has in its arsenal is the categorization of what is now termed *unity of effort* or UoE for the stakeholders involved in a conflict. With an increased focus on irregular warfare in the past decade, a need became obvious to define all elements of power and their accompanying resources by all involved in a conflict to include whole of government (WoG) participants; Department of Defense (DoD), Department of State (DoS), Department of Homeland Security (DHS), the

Intelligence Community (IC), and the whole of nation (WoN) augmentation of nonprofit organizations (NO) and nongovernmental organizations (NGOs).

Expanding this effort universally to encompass a UoE between governments, agencies, military, nonprofits, and businesses is a prerequisite for creating success in what is called a whole of world (WoW) continuity (see Figure I.1). One can see from the WoW approach diagram that the intersection defined by the Venn diagram where population needs, host-nation requirements, and the goals of the intervening forces dissect is the "zone of influence" that leads to the outcome or desired effects, or in other words, the endstate. The WoW approach lays the groundwork for approaching a collaborative vice cooperative participation by all stakeholders. Of note is that the catch phrase "whole of X" or W a oX has found its way into our decision-making lexicon. I will expound more on this in the following chapters.[10]

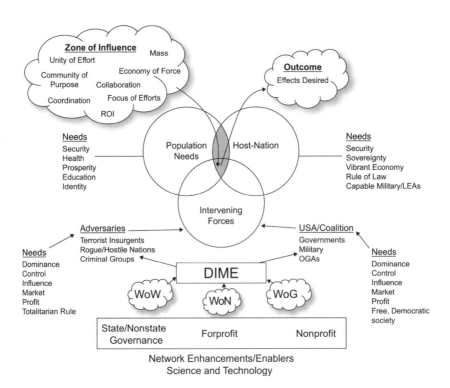

Figure I.1 Whole of World Approach

LEARNING FROM THE PAST . . . OR NOT

When the Cold War ended, hopes and expectations for a more peaceful world, eschewing violence, soared. The United States had contained the Soviet Union and then saw it through to its demise. Clearly an endstate for the termination of this 50-year war had been achieved. With the monolith known as the Soviet Union gone and the leviathan of the sole super power taking form within the United States, a new peaceful world order would undoubtedly take shape. A paroxysm of violence surely would be unlikely now that the Cold War had ended.

Instead, that hope was dashed just a few years after the wall came down with the early 1990s' deadly conflicts in Bosnia, Somalia, and other lethal troubled spots such as the genocidal-effected country of Rwanda.

> In the first twelve years of the post-Cold War (1990 to 2001), fifty-seven major armed conflicts took place in forty-five countries. In the first half of this period, the number of conflicts ranged from twenty-eight to thirty-three per year. Although the incidence of conflict dropped as the post-Cold War era stabilized, the number of conflicts has held steady since the late 1990s at around twenty-five conflicts per year.[11]

Most of these conflicts have been intrastate and only four—the Iraqi invasion of Kuwait and subsequent international intervention, the border war between Ethiopia and Eritrea, the ongoing conflict between India and Pakistan, and the U.S.-led invasion of Iraq in 2003—could be called conventional interstate conflicts.

What is interesting is that the vast majority of these conflicts were instantly successful in the opening forays. However, after the initial victories most just never ended or, if they did, they ended poorly, leaving those involved grasping for the meaning of their intervention and participation. Chapter 2 will review examples of this dynamic in greater detail.

Although the "big" wars like World Wars I and II and even Korea clearly had a direction for closure or an endstate, the smaller wars seemed to pit those involved into a foray of habitual failure. But is that really true? The research and development giant, RAND Corporation, determined that the images that populate societal thought today are failures in the conflicts in the past few decades and not the many successes over the years. The conflicts in Mozambique, Namibia, Cambodia, El Salvador, Albania, Sierra Leone, Macedonia, Bosnia, Kosovo, East Timor, and Liberia arguably have all ended on the positive (albeit somewhat) side. Today, all but Cambodia are free, to an extent, although perhaps not democratic societies.

In the last 20 years or so, the United States has had a part in building seven societies; clearly some thought existed in defining war termination and endstate for these conflicts. Therefore, the criticism that we habitually don't rebuild nations after we leave is not entirely correct. Kuwait,

Somalia, Haiti, Bosnia, Kosovo, Afghanistan, and Iraq are all examples of some levels of success in our nation's ability to rebuild a nation after conflict, and given this group—all but Haiti were Muslim—to say we do not understand how to work in that culture, as some would argue, is also not entirely correct.

"There are perhaps a finite set of ways that disputes or conflicts can come to an end. The key part is that there is always some type of political process resulting in accommodation."[12] With accommodation comes compromise and, at times, what is called *mass collaboration*.

"Whole of World mass collaboration is based on the theory that collective actions' activity through large numbers of participants and providers working autonomously, but with common purpose toward modular aspects of a single issue, can dynamically unleash greater innovation, resourcefulness, and situational awareness."[13] Some may think this a Pollyannaish theorem considering the corollary that collaboration is managed not through control or directives, but rather based on the collective consciousness of self-enforced activity of those choosing to participate and mediated by the shared understanding of an issue or threat.

Although hopeful, this mindset assumes a mass effort toward a common good, or at least a common effort to achieve something in that realm of "good." The concept goes on to convey that it is not about amassing "cooperation" since there is no negotiation and directives motivating consensus. Instead, the heart of the collaboration is based on a common "hearts and minds" purpose defined by a common issue that is communicated in global reference through all forms of mass media down to the lowest level of personal interaction. Examples of these issues include infectious diseases, global terrorism, peacekeeping, conflict prevention, education, natural disaster prevention or mitigation, human rights and rules of social law, illegal drugs, etc. Theorists who support this concept believe that without a common purpose expressed effectively, mass contribution will never amount to anything along a focused cause to "direct" efforts.

It is clear that there was a level of success for each of the conflicts in the waning years of the 20th century and the first decade of the 21st century; however, was the outcome optimal and supportive of the objectives defined at the outset? Did the leaders involved take a myopic nation-focused approach in setting goals or did they look toward a WoW cooperative collaborative process for defining an endstate?

In 1997, I presented an Institute of Land Warfare paper, "Terminating the Ground War in the Persian Gulf: A Clausewitzian Examination," for the Association of the United States Army that showed the overarching goal for the combatant forces in Operation Desert Storm was to expel the Iraqi occupation of Kuwait. That goal was achieved in the first 100 hours of the ground war. We had achieved what we planned from the beginning of Desert Shield, which started some eight months prior. The ensuing arguments that questioned if we should have exploited this success and continued to defeat the Iraqi army in detail by "going all the way to Baghdad" are moot because the seizing of the Iraqi capital was never part of the original plan. Just as important, the condition did not change on the ground or in the operating environment at the time to warrant a change on what is commonly referred to as *mission creep*.[14]

The limited objective of thwarting Saddam from Kuwait is an example of success in terms of an exit strategy and perhaps should be used as a model for others to follow. However, one could argue that the establishment of goals and amassing consensus from all the stakeholders involved during Desert Shield and Desert Storm were aided by the fact that a true coalition was formed from the onset (see Table I.1).[15] When there is no outside influence to impact the decision-making processes by the two opposing belligerents, it is commonly referred to a *rational model* process. The rational model implies that states in wars are uni-actors and serve as the central decision makers. It implies that both belligerents understand the value of the means and ends of the war and possess the knowledge to calculate the relative power and value of the other side's means and ends. Lastly, the model implies that both sides can calculate the cost involved in attaining the ends.[16]

In even the simplest of conflicts, the rational model is arguably overly simplified. Nonetheless, history shows that a coalition that establishes universal goals and objectives via a loose alliance of actors struggling to gain a unity of effort is more apt to achieve a successful war termination and conflict resolution set of objectives. For the rational model discussion, the alliance of the willing (a la Operation Desert Shield and Operation Desert Storm) are seen as one as they vie against an opposing belligerent. Although the quantity of the assembled members of an alliance or coalition is not important (see Table I.2), the association with the conflict is what is critical. For more on this examination, see Chapter 3.

Another example of going beyond the guns and steel is in exploring what happened to the axis powers, Germany and Japan, after World War II. The defeat of Germany and its postconflict resolution in World War II was arguably a great success. Some would argue that Nazi Germany was a Western-type nation-state before its fall, so it was easier to rebuild. The same cannot be said for Japan, which can be labeled an imperialistic nation-state.[17] However, in the same era its reconstruction led to an even greater success; especially in its economy. How does one explain this phenomenon? We will explore this more in Chapter 3.

Many of the conflicts listed and discussed in the preceding paragraphs and throughout this book have had acceptable outcomes, but at what cost? Did waiting too long before deciding what to do after the bullets stop flying result in a greater cost in both resources (money, time, equipment) and lives? Did the decision-making lag impact the overall outcome militarily,

Table I.1
Desert Storm Theater Objectives and Phases

Theater Objective	Phase I: Strategic Air Campaign	Phase II: Kuwait Theater of Operations Air Supremacy	Phase III: Battlefield Preparation	Phase IV: Ground Offensive
Disrupt leadership and command and control	X			
Achieve air supremacy	X	X	X	
Cut supply lines	X	X	X	X
Destroy Nuclear, Biological, and Chemcial capability	X		X	
Destroy Republican Guard	X			X
Liberate Kuwait City				X

[1] Grange, Brig. Gen. David L. and Swanson, Scott, "Confronting Irregular Challenges in the 21st Century," Irregular Warfare Concept Series: Whole of World Collaboration (March 10, 2009), 2.

Source: CENTCOM operations plan, December 16, 1990[1]

Table I.2
List of Coalition Forces by Number of Military Personnel

Country	Number of Personnel	Comments/Major Events
United States	575,000–697,000	Operation Desert Shield Battle of Khafji Battle of 73 Easting Battle of Al Busayyah Battle of Phase Line Bullet Battle of Medina Ridge Battle of Wadi Al-Batin Battle of Norfolk Operation Desert Storm
Saudi Arabia	52,000–100,000	Operation Desert Shield Battle of Khafji Operation Desert Storm
United Kingdom	43,000–45,400	Operation Desert Shield Operation Granby Operation Desert Storm
Egypt	33,600–35,000	Operation Desert Storm
France	18,000	Opération Daguet
Syria	14,500	Operation Desert Storm
Morocco	13,000	Security Personnel
Kuwait	9,900	Invasion of Kuwait Operation Desert Storm
Oman	6,300	Operation Desert Storm
Pakistan	4,900–5,500	Backup team
United Arab Emirates	4,300	Operation Desert Storm
Canada	2,700	Operation Friction
Qatar	2,600	Battle of Khafji

Bangladesh	2,200	(Operation Moru-prantar)
		  Security Personnel
		  including 2 field
		  ambulance teams

Italy	1,200	Deployed Panavia Tornado strike attack aircraft
Australia	700	Australian contribution to the 1991 Gulf War

Netherlands	700	Naval deployment; Air
		  Force deployments of
		  surface-to-air missiles to
		  Turkey and Israel

Niger	600	Patroller group
Sweden	525	Field hospital
Senegal	500	Base Guards

(continued)

**Table I.2
(Continued)**

Spain	500 on the field/ 3,000 off the coast	Engineers
Bahrain	400	Base Guards
Belgium	400	Base Engineers
Poland	319	Operation Simoom Naval and medical deployment
South Korea	314	Medical and transportation support
Czechoslovakia	200	Operation Desert Shield Operation Desert Storm Czechoslovakia in the Gulf War
Greece	200	Pilots
Denmark	100	
New Zealand	100	2 C-130 Hercules transporter aircraft
Hungary	50	
Norway	50	Naval and medical deployment

socially, politically, and economically? Without much analysis, one can conclude that not planning for war termination and conflict resolution at the onset of conflict had a significant impact in the history of war. The question is, how much of an impact? Based on the answers to this question, how determined should decision makers be in the future to ensure the best possible outcome of conflicts yet to occur?

I will show that outcomes based on exit strategies, or the lack thereof, truly fall in three categories: optimal, acceptable, and unacceptable. It is clear that these assessments have to be placed against some definitive goals; a shortfall already cited in our strategic postconflict planning. Obviously, once the goal is defined, we should strive to achieve as much of an optimal solution as possible. Even if we believe the objective or goal has been achieved, then the question becomes, through whose lens is it viewed? For more on this subject, see Chapter 7 where a war termination and conflict resolution methodology that takes into account termination categories and conflict resolution potentialities is discussed in detail.

IT HAS BEEN DONE

Those with power to start a war frequently come to discover that they lack the power to stop it.[18] In part, governments tend to lose sight of the ending of wars and the nation's interests that lie beyond it, precisely because fighting a war is an effort of such vast magnitude.[19] However, defining the ending of conflict is indeed possible. Clearly, history is replete with examples. Analysts, historians, and planners have filled volumes attempting to categorize war termination. In *Conflict Termination: A Rational Model* (1992), Bruce B.G. Clarke states that ". . . to understand how to terminate a conflict one needs to first understand the sources of the specific conflict."[20] In deciding whether or not to initiate hostilities, statesmen may attempt to weigh the risks and costs of avoiding war, on the one hand, against the dangers and possible gains of a war, on the other.[21]

With this premise in mind, Clarke and William O. Staudenmaier's "Conflict Termination in the Third World: Theory and Practice," define six major ways to categorize the ending of conflict or its termination. Recognizing the genesis behind a fight and then how it ensued arguably helps strategists and planners to define the outcome. The following categories are typically agreed on by most planners and policy makers as *war termination* categories:[22]

1. Armistice, truces, and cease-fires
2. Formal peace treaties
3. Joint political agreements
4. Declaration of a unilateral victory by victor
5. Capitulation by the loser
6. Withdrawal of one of the parties.

The war termination categories are the first part of an algorithm I will define in my strategy for ending wars once they begin and, in some cases, prevention (covered later in a separate category). The identification of these categories eventually drives the determination of war termination in a conflict. I will explain these categories in greater detail using vignettes so the reader can better understand each as it relates to why a war began and its subsequent outcome in the past in relation to those conflicts of the future. A more precise understanding of each can be found throughout the remainder of this work starting in Chapter 1.

The other half of the algorithm to bring about an endstate is *conflict resolution*. Once a war is brought to a conclusion, resolution of the conflict becomes the necessity. War termination and conflict resolution, although separate entities in work effort, depend on each other to reach a satisfactory endstate or goal or, even more pronounced, a posthostility aim. The conflict resolution categories that strategists and civilian military leaders find themselves striving for—unfortunately in the nth hour—include:

• Nation building
• Stability, security, transition, and reconstruction operations

- Economic development
- Humanitarian relief
- Transitioning security to the indigenous force
- Establishing a democratic nation
- Countering an insurgency and maintaining a lasting presence
- Invading a country to establish an imperial footprint or continual military presence
- Providing persistent foreign internal defense
- Withdrawing under fire without a deliberate transition plan.

Note: a combination or inclusion of any of these categories with the other(s) is acceptable if not optimal in some cases. These categories define what is commonly known in military jargon as an *endstate* or as stated earlier a *posthostility aim*. It is my premise that they are merely part of the equation that defines the ultimate outcome of war.

Although these categories guide the reader through the war termination and conflict resolution processes that this work supports, historically these categories are lacking in two ways. First, they are reactive in their results. According to analysts like Clarke, conflicts unfold in a phased process. In his model, decision makers decide during the conflict not only when to move on to the next phase, but also how the next phase is characterized. It is my argument that if a plan for war termination and conflict resolution are in place prior to the conflict then decision points can be defined and conditions established prior to the fight to help guide its direction. If measurable decision points are needed, they can be developed to help guide decision makers.

Second, the war termination categories tend to describe "how" the conflict will end as opposed to what happens after the fight. It is great to devise, say, a peace treaty or cease-fire between warring factions, but defining how to stop the fighting is not enough. Determining the conflict resolution is just as important as the war termination in this effort to achieve a desired endstate.

Therefore, the penultimate portion of the algorithm is to define the goal for conflict resolution based on the war termination precepts. The premise, therefore, for this work is that *an endstate is achieved when* **objectives** *are met,* **war termination** *is defined, and* **conflict resolution** *is identified and resourced*. This is the algorithm of sorts used eventually to define my methodology. Later in this work we will review the categories for each portion of the algorithm and prescribe a methodology for identifying each portion to eventually describe what happens thereafter.

Defining the bins from which conflicts are viewed is helpful to see which conflict has been driven to an end by proactive engagement and which is destined to end out of inertia, inactivity, and ultimately by default. Strategists are forever attempting to drive results by engaging with policy makers and by influencing public policy. At this point, it is instructive to introduce a basic (albeit, somewhat sophisticated) understanding of the strategy-policy paradigm. Understanding the difference between strategy, grand strategy, and policy assists the reader in understanding the most important and, often the most constraining, element of the employment of a nation's power—its resources.

To understand how money is spent, equipment is bought and employed, and people are assembled to accomplish the full array of tasks associated with all phases of conflict, one must first understand how the resources are managed within the hierarchy of those that have a stake in a conflict. In short, to understand how to devise a conflict exit strategy, one first has to comprehend what it means to define a nation's strategy as a framework and how that strategy is nested with a nation's policy. This concept will be covered in some detail in Chapter 2.

WHY THIS BOOK?

Some would question why yet another book on strategy? History teems with examples of how and why a nation decides to deploy a force to combat; but rarely does it have a definitive plan to extricate itself from the fray. In this text, I demonstrate that clear goals coupled with resources sufficient to accomplish these goals can maintain purpose and guide the fight through transitions from high-intensity conflict to low-intensity conflict, counterinsurgency, and reconstruction. As a result, these transitions can be predicted, planned, and resourced in advance, preventing wasted resources and operations as the stage of conflict is identified by trial and error, as we have seen in recent operations in the Middle East, both in Iraq and Afghanistan.

One may think that having a plan for conflict resolution is intuitively obvious for policy makers. Inherently decision makers, at least those with a great amount of experience, strive to define a definitive endstate for any strategic, operational, and especially a tactical mission. However, a quick search and analysis of existing books, articles, and manuscripts shows that there is indeed a void of written works that provides not only recognition of this strategic abyss, but also a lack of methodology for achieving success beyond what is typically termed *victory*. This text is my effort to show that most conflicts in history have begun with little thought toward developing a definitive plan for resolution. An exit strategy, or a war termination strategy and its accompanying conflict resolution strategy, I argue, have been the greatest void in devising plans for conducting combat operations in the recent past. The two are not mutually exclusive and have to be considered in tandem to achieve success.

Although dozens of books on the market today explore conflict resolution in some form, most focus on deciding when and if to fight. With the current world situation in mind, one is left wondering what happens when, for better or for worse, a nation has already embarked on a path to war. Beyond questions of *whether* to fight, beyond the analysis of success and failure *during* a fight, lie queries of "When is the fight over? Have the nation's objectives and goals been met? How do we return to a status of peace or to, at least, prewar conditions, or perhaps even better?" As stated earlier, these questions have proven difficult for policy makers and policy analysts alike. Analysis of justice in war proves an easier path. In retrospect, the more difficult, but equally important, task lies in the analysis of events leading up to war and the desired endstate shaped by the original intentions.

Furthermore, most literature that exists on this topic deals with terminating major wars, and very little has been written about conflicts that stop short of conventional fighting. Accounts such as *The Generals' War* by Michael R. Gordon and General Bernard E. Trainor and *Crusade* by Rick Atkinson briefly examine the issue of exit strategies. Though these works may provide insight into the political machinery of war, they do not shed much light on the difficult process of developing a conflict exit strategy.

Fred Charles Iklé's *Every War Must End* is perhaps the most in-depth view on war termination in print. However, it focuses mostly on how to stop a war once a nation is involved. Iklé covers planning for war termination before it starts, but mostly in the form of nuclear deterrence—an obvious but waning concern for most of the world and its strategists since the end of the Cold War. Of all the published works that exist, however, *Every War Must End* provides an excellent analysis of external issues that affect termination: the fog of military estimates, nuclear weapons, and political objectives. Originally published in 1971, Iklé updated it in 1991 after Operation Desert Storm and then again in 2005 in an attempt to capture lessons from OIF and OEF. Coupled with Geoffrey Blainey's *The Causes of War*, one can garner a great understanding of why wars start and how they came to end; however, it doesn't provide a strategic meaning for why the goal for each war was sought; in fact statements like ". . . I thought that the people responsible [for Desert Storm] ought to start thinking about how it would end . . ." is a common theme in this work as in most that explore postconflict operations.[23] In short, not having a plan in place to terminate war and acknowledge the activities required to resolve the conflict is akin to telling a driver not only where it is he went after he got there, but to calling him while he is going to tell him to stop as opposed to giving him destination and a strip map in the first place.

The collection of essays in *Conflict Termination and Military Strategy* by Stephen J. Cimbala and Keith A. Dunn identifies how termination goals affect military strategy. However, like most books currently in print, these were published pre-9/11 and do not cover conceptually and in practice the ensuing effort to defeat terrorism. In fact, the vast majority of the work on war termination and conflict resolution date back to the Cold War with central themes that concentrate on major conflicts and the resulting super-power interaction. There is little doubt that the shifts in warfare since the end of the Cold War demand an all-inclusive review of this keen part of war planning lest nations squander their resources as they meander through conflicts without guidelines for termination.

In *Through a Glass Darkly: Looking at Conflict Prevention, Management, and Termination,* Cimbala shows why the prevention, management, and conclusion of war all require an understanding of the subjective aspects of decision making as well as the hardware and tactics of military operations. He reviews past cases of U.S. security policy decision making and provides a preview of some future problems and how they are combined to distill important lessons about coping with conflict in the post-Cold War world. Again, although a noble effort, it is an explanation of the accident after it happens as opposed to attempting to form some strategy to prevent an unintended outcome or settling for a suboptimal solution.

Bruce C. Bade's essay "War Termination: Why Don't We Plan for It?" gives the most comprehensive account of the actual problems associated with war termination. As the title indicates, the 28-page essay examines the military's reluctance to plan for war termination and highlights the reasons why. Bade faults the U.S. mentality by claiming that Americans like to think that war termination will take care of itself. He correctly identified the nation's shortfall in determining war termination as ". . . the details of how the vanquished will be managed following capitulation of the enemy and cessation of hostilities" and correctly defines *war termination* as a process rather than an event which by definition requires planning.[24] However, he doesn't address *conflict resolution* as an added entity for devising a plan for a coherent endstate.

Bade also does well in defining that the Clausewitzian point of culmination, although it has ". . . the potential as the focus of a planning framework for war termination,"[25] it is, however, nothing more than a point of equilibrium between the attack and the defense where the offense can proceed no further without becoming relatively weaker than the defender. In short, as Bade so correctly points out, the *point of culmination* is a theoretical premise used to define when to stop to perhaps negotiate a peace, as opposed to planning those measures in advance and, therefore, only part of the overall war termination plan.

The majority of Bade's essay, however, is reflective of the faults of our strategy and only a portion speaks to "rational frameworks" for devising phases of conflict from identification of the dispute through settlement. Postulating that in the "final analysis, all disputes will end through some type of political process that will result in a political accommodation that has some minimal amount of formal structure," Bade understood that the human factors play a large part in defining a successful outcome in a conflict. Of great benefit is his listing of "types" of termination that we will explore later on in this work.[26]

Finally, Clarke's "Conflict Termination," previously discussed in this introduction, provides an explanation of "the interrelationship between the mechanisms for ending a conflict and the nature of a conflict."[27] Clarke's model is applicable to understanding an analytical process for terminating a conflict. I refer to this model throughout this work as an adjunct to the analysis and framework I develop.

In the end, with the ongoing wars in Iraq and Afghanistan, the need for a book on terminating conflict and bringing war to a resolution is obviously needed. The impact on the fighting force has been great. As of July 2009, 73 percent of company-grade officers and 83 percent of field-grade officers were either pending deployment or had spent at least two months deployed in the previous five years. With these types of statistics, we owe it to our fighting forces, their families, their country, and those affected by their professional way of waging war to have defined exit strategies and resolutions for conflict.[28]

ON WRITING THIS BOOK

Beyond Guns and Steel: A War Termination Strategy expounds on historical examples to show that our national command authority habitually ventures into combat lacking clear, concise goals and objectives and then conducts reactive "crises management" in a disorganized stumble from conflict to shaky peace. Our current doctrine ends with the conduct of war. Even the newest field manual, *FM 3-24: Counterinsurgency* ends with the defeat of the insurgent; begging the question, "What's next?".

Clearly, there is a void in the doctrine and the policy that thrives on tethering itself to a well-thought-out blue print for success. This book gives a snapshot of conflicts like today's ongoing wars in Iraq and Afghanistan. It also explores many other conflicts, big and small, whose histories have been penned; however, this work explores that period beyond war after the main threat had been defeated, but the path to normality remains unclear.

As you will see in this book, the constraining element besides an innate inability to "see the forest through the woods," is resourcing of those agencies called on and required to support a conflict.[29] Former Secretary of State Condoleezza Rice addressed this need in a speech at Georgetown University, stating that we must focus our energies *beyond the guns and steel* of the military, beyond just our brave soldiers, sailors, marines, and airmen. We must also focus our energies on the other elements of national power that will be so crucial in the coming years.[30]

The DoD habitually is funded to accomplish its combat role as the supplier of *hard power.* But as the wars in Iraq and Afghanistan have shown,

there is a fine line between what tasks the military are required to perform to win the battle and the effort to transition the fight to the civilian
authorities and the indigenous popularly elected government. Both OIF
and OEF have established a need for this war termination-conflict resolution transition period. In fact, 2010 was a pinnacle year for the war in Iraq
where the Commander of U.S. Forces, General Raymond T. Odierno and
the U.S. Ambassador to Iraq, Chris Hill, developed a comprehensive joint
campaign plan that outlines the transition of military activities to those
in the embassy, the activities requiring transition from the U.S. military
to the government of Iraq and the Iraqi Security Forces, and those that
either need to transition to a contracted element or fade away for a lack of
requirements at this point in the fight. Although these transitions at first
seem plausible and then achievable, unfortunately, the *soft power* providers
like DoS, DoT, DHS, various rule of law agencies, IC organizations, NGOs,
and other stakeholders (see Chapter 5) are usually under-resourced.

As stated earlier, massing resources and applying UoE to a WoG,
WoN, and even a WoW problem can create "positional advantage" for
the United States and its allies, and, where applied appropriately, can be
used against adversaries or other national security threats. As I will show
later, adversaries are currently using similar approaches to gain support
for their causes.[31] Ensuring a resourced, well-developed lifecycle plan that
incorporates the nullification of an enemy set, ensuring that the indigenous force and government is capable and has the capacity to sustain
itself, and then tight-wiring a resourced transition force replete with the
skill sets and talent pool (to include the stamina) are essential in achieving
a suitable endstate.

In a speech at Kansas State University on November 26, 2007, Secretary
of Defense Robert M. Gates said that, "After all, civilian participation is
both necessary to making military operations successful and to relieving
stress on the armed forces who have habitually led the resourcing, manning especially, in most fights." Regardless of improvements in effort or
focus, there is no substitute for financial resources. "Funding for nonmilitary foreign-affairs programs has increased since 2001, but it remains dis-
proportionately small relative to what we spend on the military and to the
importance of such capabilities."[32]

Gates also stated that, "Despite new hires there are only 6,600 professional Foreign Service officers (FSOs)—less than the manning for one aircraft carrier strike group." In a country that spends only about 4 percent of
its gross domestic product on defense and only a fraction of that on FSOs,
it is clear that we continue to be a military at war focused on finishing the
wars it starts versus a nation focused effort.[33]

But again, it seems in an effort to fix what is broken after the conflict,
war termination conundrum leaders always try to "throw more stuff at
it." New organizations and restructuring for success are undoubtedly part

of the solution to fix the problem. However, we must first correctly define the problem.

As stated, this study does not judge the reasons why the United States decided to enter a conflict; however, it does examine how the rationale for intervention influences the exit strategy. It doesn't develop a "one size fits all" model to churn out solutions to finish conflicts once they start, for that would be futile and, at the very least, suboptimal. Nor does it ignore the fact that the changing face of battle as it unfolds can make any plan, regardless of the thought put into it, of no value if the conditions change.

What this study does provide is a determination of the commander's and his staff's role in war termination planning as part of the initial planning process when the decision to go to war is made. Specifically, this work concentrates on the difficulties commanders and decision makers face when translating the political objectives into a military termination strategy and conflict resolution framework and what results at the strategy and policy level are desired. This book also explores the potential need of meeting operational and strategic objectives established at the outset of the campaign and adjusting them as the environment changes as the fight ensues.

From 1492 to 1941, arguably the first 500 years of the American dream saw the founding of our land and the demise of its biggest threat. Unlike the Europeans of the 20th century, Americans dominated themselves on the continental United States before attempting to dominate the rest of the world.[34] It is the first time in half a century that power didn't reside in Europe. By 1991, the sole superpower was the United States and the center of the international system emanated from both of the U.S. borders.

Regardless of how we struggle to bring a war to its end, the United States has done allright by any standard in history. However, there is little doubt that the experiences in places like Beirut, Somalia, Haiti, the Balkans, Iraq, and Afghanistan, to name just a few, have left nations, and in particular the United States, struggling to achieve an acceptable endstate which, in turn, not only makes one look to find a solution but also makes an effort to establish a framework for future conflicts. This struggle has left strategists and policy makers reliant on coherent strategic and operational plans that are reactive instead of proactive in their efforts.

CHAPTER 1

The Fog of Postwar

Pure military skill is not enough. A full spectrum of military, paramilitary, and civil action must be blended to produce success. . . . To win in this struggle [we] must understand and combine the political, economic, and civil actions with skilled military efforts in the execution of the mission.

—President John F. Kennedy
Letter to the United States Army, April 11, 1962

In August 2009, a year after their war that left Russian forces occupying the seceding Georgian regions of Abkhazia and South Ossetia, Russia and Georgia were locked in a battle of words, images, and propaganda over how the world will remember the conflict. Each sought to be exonerated, each wanting its adversary condemned—and both proclaimed that the other side was spoiling for a new fight. The fog of postwar was nearly as thick as when the fighting was raging. There was even disagreement on what day marks the beginning of the five-day conflict.[1]

Although the crisis showed Moscow is willing to resort to force when it feels it is necessary, the war revealed the shortsightedness of its policies and ultimate aims. The Kremlin seems to have had no plan for what to do with the "statelets" it has adopted or how to move forward in its relations with Georgia. There is no grand strategy nor is there a master plan.

Russia still has very little soft power, beyond the military power—which is considerable—beyond the energy power, there is less soft power and that means that fewer people are going to be persuaded to what it is that Russia represents, even in the neighborhood. . . . Edward Lucas, author of *The New Cold War* and the Central and Eastern Europe correspondent for "The Economist," agrees that if the Kremlin intended to send a signal to the capitals of what it calls its "near abroad," it may have miscalculated. "Russia is now just beginning to realize that being feared and distrusted by your neighbors isn't necessarily the basis for getting what you want in policy."[2]

This is clearly a case of a miscalculation in understanding what is the endstate for the Russian version of Lebensraum, leaving all involved confused about the outcome; clearly an ambiguous fog of postwar.

At this point in the book, we examine the trials and tribulations and eventual outcomes of conflict at the point where the fighting stops and conflict resolution begins; a time in space that is better thought of as an overlapping continuum as opposed to a discrete point. In a play of words stemming from the Clausewitzian *fog of war*, I call this the *fog of postwar*. In short, the fog of postwar speaks to the ambiguities of when war terminates and when conflict resolution takes place. It is not just about the time the fighting comes to fruition. An added level of uncertainty is what occurs to allow war to end and what types of activities are required to begin the reconciliation part of war termination to recapture the peace or define the newer postwar environment. This, as stated in the introduction, is conflict resolution.

To refresh for those unfamiliar with his theories, the great Prussian military analyst Carl von Clausewitz wrote:

> The great uncertainty of all data in war is a peculiar difficulty, because all action must, to a certain extent, be planned in a mere twilight, which in addition not infrequently—like the effect of a fog or moonshine—gives to things exaggerated dimensions and unnatural appearance. What this feeble light leaves indistinct to the sight, talent must discover, or must be left to chance.[3]

The fog of war is Clausewitz's way of describing the opaqueness and resultant uncertainty inherent in any military campaign. Military planners must take into consideration that conditions on the ground will vary from war to war—and even from battle to battle—and thus not assume that strategies that work in one situation will necessarily translate to another. For instance, to assume that a strategy that has worked in Iraq (the surge, for instance) will certainly work in Afghanistan—with its much different terrain, familiarity with central government, history with occupations, educational levels, economic capacity and potentiality, and culture— would be viewed by Clausewitz as naïve at best and foolish at worst.[4] But not to have a plan, Clausewitz and any other strategist would argue, is as equally folly.

As discussed in the Introduction, this chapter goes into greater detail in exploring the query: When does war come to fruition and does it ever really terminate? Instead, if nothing is done to prevent the killing, is there a point of culmination where both sides "die on the vine" or exhaust themselves to a point of incapacitation? Or do we habitually stumble into a foray of reconstruction and stabilization efforts and the injection of civil response elements to nation-build after a level of destruction occurs through kinetic operations? The answers lie in how prepared is a military, a nation, and the global community for the initial fight; and more importantly, how prepared are they in their plan to transition to the different phases that ensure success after the initial bullet is fired? Or perhaps, even more importantly, what processes are in place so that planning for such transitions is natural and, therefore, resourced and not just an afterthought?

FIRST, WE START THE WAR . . .

The vast majority of international relations theory and conflict research has concentrated on conflict prediction, with a view to crisis prediction, conflict causation, and prevention. A vast amount of literature has been assembled on this, mostly framed by the Cold War context of the 1960s. However, the results have often been somewhat lacking, oftentimes stating the obvious, or providing description functions too weak to deal with a system where both false-positives and false-negatives have very high ancillary costs.[5]

In contrast, this work is focused on endstates, with a view of not necessarily how they are won but how they come to an acceptable (preferably optimal) conclusion. It is instructive at this point to discuss the *rational expectations* theory. This is a theory of behavior initially drawn from economics, but used widely elsewhere like in game theory. It refers to a theory of behavior that states "actors will take choices designed to maximize their perceived payoff according to their schemes of utility." This can be simple re-phrased as "people will act in what they think is their own self-interest."[6]

This is important because if states engage in conflict for pure selfish reasons, then the determination of war termination and conflict resolution is based on those specific motivations and any determination using the previously discussed rational model where a desired coalesced outcome is prevalent based on a coalition mindset.

Insofar as it relates to war, rational expectations theory makes the following predictions:

- States become involved in wars because they expect the benefits of fighting to outweigh the costs.
- States will continue to fight so long as their expected benefits from fighting exceed their expected costs.
- Warfare will continue until all sides see it as being in their interest to stop.[7]

As Alistair Morley of the United Kingdom's Defense Science and Technology Laboratory points out, the expected benefit exceeding expected cost does not imply an "expected net gain" from the conflict; settlements are often merely the "least bad option" for the warring parties. He goes on to state that warfare tends to be, in economic terms, a negative sum game; both sides can end up worse than before. Where states derive a net gain from a successful war, the losses of the losers tend to be greater than the winners' gains. This is not exactly surprising; little is more destructive of human utility than the expenditure of resources with the intent to destroy other resources—an obvious byproduct, if not a deliberate intention of warfare.[8]

As we explore stakeholders in the strategic decision-making process, it is instructive to look at the policy makers and how they are viewed in this concept:

- The policy elites (governments of various types) are generally rational within the constraints of information provided to them;
- That policy elites are generally capable of effectively controlling the actions of their owns states;
- States seek to maximize their utility according to a consistent scheme;[9]
- The use of actual costs, or what the state has suffered or gained from fighting thus far[10] and outstanding benefits, or the remaining war aims the state could potentially fulfill by continued fighting[11] as a proxy[12] for expected costs and benefits;
- The problem of "censured" data is not overly distorting.

Understanding the players or stakeholders gives the strategist and planner a sense of whom they are dealing with. Just like other resources—time and money—the people involved in the conflict and what they desire is an important variable when devising war termination and conflict resolution strategies.

ONCE IT STARTS IT CAN'T STOP. . . .

Much like Newton's First Law of Motion, which states that a body in motion will remain in motion in a straight line unless acted on by an outside force, the same holds true for war. Warring parties will remain at war unless an outside force intervenes or one side capitulates or is annihilated; which is obviously a form of war termination albeit not one the losing side would choose. What complicates matters is that as the fighting continues, its purpose often changes, thus making the ability to define war termination and conflict resolution exponentially more difficult. More appropriately, Iklés says in *Every War Must End* that ". . . fighting often continues long past the point where a rational calculation would indicate the war should be ended."[13] In fact, this line from Iklés is what ran through General Colin Powell's mind on February 27, 1991, when he recommended to President George Bush to stop the fighting against the Iraqis in Operation Desert Storm.[14]

Powell knew that the objective of pushing the Iraqis out of Kuwait had been achieved. To let the U.S.-led coalition proceed further and "on to Baghdad" as some suggested and which Powell opposed was not in the set parameters of the endstate:

- Effect the immediate, complete, and unconditional withdrawal of all Iraqi forces from Kuwait;
- Restore Kuwait's legitimate government;
- Ensure the security and stability of Saudi Arabia and other Persian Gulf nations; and
- Ensure the safety of American citizens abroad.[15]

As stated, to go beyond expelling Saddam's forces from Kuwaiti borders was not part of the original plan and the conditions didn't exist for a full-fledged invasion of Iraq and seizing its capital. Most importantly, there was NO plan for the aftermath of such an extension. No resources were committed to the eventual need to counter a rising insurgency as already seen by the Shi'a uprising in the south, and there was certainly no plan to address the effort seen now in Iraq for establishing essential resources and a legitimate government (more on this in Chapter 3).[16]

Although no plan survives the initial employment or movement of forces because the changing in conditions habitually causes changes to ultimate objectives, the conditions in March 1991 did not allow for an all-out invasion of Iraq and its capital, Baghdad. By overextending the offensive, it would have most certainly called for a protracted war and the challenge for resourcing of postconflict resolution strategy.

One of the best explanations in defining how the policy makers and strategists make decisions comes from Harry R. Yager's *Strategic Theory for the 21st Century: The Little Book on Big Strategy*:

> The strategic process is all about *how* (way or concept) leadership will use the *power* (means or resources) available to the state to exercise control over sets of circumstances and geographic locations to achieve *objectives* (ends) in accordance with state policy. Strategy provides direction for the coercive or persuasive use of this power to achieve specified objectives. This direction is by nature proactive, but it is not predictive.[17]

This is important to understand, for later in the text I will show that although I am recommending an algorithm of sorts, my premise is that it is essential to define a war termination strategy in concert with a conflict resolution method in the planning process before conflict begins or is initiated so that an endstate can be achieved when specified objectives are met.

Strategy assumes that while the future cannot be predicted, the strategic environment can be studied, assessed, and, to varying degrees, anticipated and manipulated. Only with proper analysis can trends, issues, opportunities, and threats be identified, influenced, and shaped through what the state chooses to do or not do. Thus, good strategy seeks to influence and shape the future environment as opposed to merely reacting to it. Strategy is not crisis management. It is to a large degree its antithesis. Crisis management occurs when there is no strategy or the strategy fails to properly anticipate current practice stemming as seen in the failed fight in Somalia in 1993. Since then leaders have been conducting *consequence management* which, in my view, is another term for reacting expeditiously to poor planning. Thus, the first premise of a theory of strategy is that strategy is proactive and anticipatory, but not predictive.[18] And so it is; consequence or crisis management equates to failure according to Yager and is part of the premise of this work.

Most books on this topic first explore how the fight began. Many talk about ending wars but doing so presupposes that the participants are stuck in a fight and looking for a way out. Still others draw a distinction between war termination and war resolution; the former focusing on stopping the violence and the later honing in on what to do with the nation and its people once the fighting stops.

Many strategists and planners have viewed defining endstate categories as important. This book discusses war terminating and conflict resolution as a dependent pair requiring a plan to achieve both in unison in advance of the fight. Clearly, history has shown it to be so. Analysts, historians, and planners have studied it in great detail and there are volumes of works published that attempt to categorize areas for war termination. So, as we venture into defining how this "dependent pair" interacts to bring us a desirable endstate, we must first define each in categorical terms.

In Bruce B.G. Clarke's model for war termination as drawn from his 1992 pamphlet, *Conflict Termination: A Rational Model,* he defines six major ways that disputes or conflicts can end.[19] These six categories, while not exhaustive, provide the framework for the strategy provided in this work.

1. Armistice, truces, and cease-fires
2. Formal peace treaties
3. Joint published agreements
4. Declaration of a unilateral victory by the victor
5. Capitulation by the loser
6. Withdrawal of one of the parties

ARMISTICE, TRUCES, AND CEASE-FIRES

Historically, there has been confusion about the precise meaning of each of these terms. In short, an *armistice* is an effort to take steps to demilitarize the situation. In essence, under an armistice an agreement is made by all sides to halt the fighting; no one is deemed to surrender to another. It is not necessarily the end of a war, but may be just a cessation of hostilities while an attempt is made to negotiate a lasting peace. It is derived from the Latin *arma*, meaning weapons and *statium*, meaning stopping. Usually, an armistice is a prelude to actual peace negotiations and, therefore, a sense of permanence accompanies it. It signals that one side agrees to declare the other side the "winner." The "loser" is usually given a set of conditions by which they must abide or risk resumption of warfare. A good example is in World War I, where the fighting ended, but Germany did not really surrender to the Allies, which arguably was part of the catalyst for World War II.

A temporary suspension of active hostilities is commonly referred to as a *truce*. A truce is kind of a "time out." The difference between an armistice and a truce is that an armistice is more like "game over" where a truce is a break much like a "time out" in sports.

A *cease-fire* simply means that the shooting (or violence) stops. An order is given on one or both sides to stop the fighting. A cease-fire is exactly that—all sides agree to disengage at a set date/time usually for a specific period. It does NOT mean a cessation of hostilities (a truce). North and South Korea are technically under a cease-fire, not an armistice or peace treaty. Both sides are legally in a state of war. A cease-fire is temporary.

Traditionally, if an enemy asks for a cease-fire during what it calls *peace negotiations*, it is merely a ploy for them to avoid hostile fire while they resupply their troops and replace the dead and wounded. When they are ready, they start fighting. When the other side fires back, they accuse them of violating the cease-fire and storm out of the peace negotiations. A recent example of this is the 2007 declaration made by the Shi'a nationalist cleric Muqtada al-Sadr in Iraq when he suspended the activities of his powerful Mehdi Army militia for initially six months after clashes in the holy city of Kerbala killed 52 people and forced hundreds of thousands of pilgrims to flee. As stated at the time of the cease-fire, Sadr's spokesman Sheikh Hazim al-Araji said that the aim was to "rehabilitate" the militia, which was divided into factions.[20] Although the cease-fire has been challenged a number of times by splinter Sadr elements, at the time of this writing it remains a vital part of the operating environment in Iraq so that the United States has been able to initiate the final stages of its responsible drawdown.

Expounding on Ending Conflict

It is instructive to show examples of each of these three categories to put them in context to better understand the first part of our equation: war termination. An armistice is a *modus vivendi*, or an agreement to agree to disagree; militarily speaking it is an agreement to disengage and is not the same as a peace treaty, which may take months or even years to agree on. As stated the 1953 Korean War armistice[21] is a major example of an armistice that was not followed by a peace treaty.

The United Nations (UN) Security Council often imposes or tries to impose cease-fire resolutions on parties in modern conflicts. Armistices are always negotiated between the parties themselves and are thus generally seen as more binding than nonmandatory UN cease-fire resolutions in modern international law. The key aspect in an armistice is the fact that "all fighting ends with no one surrendering." This is in contrast to an unconditional surrender, which is surrender without conditions, except for those provided by international law.

The most notable armistice, and the one which is still meant when people in Europe say simply "The Armistice," is the armistice at the end of World War I, on November 11, 1918, signed near Compiègne, France, and effective at the "eleventh hour of the eleventh day of the eleventh month."[22]

Armistice Day is still celebrated in many countries on the anniversary of that armistice; alternatively November 11, or a Sunday near to it, may still be observed as a Remembrance Day.[23] In the United States, November 11 is observed as Veterans' Day.

Other armistices in history include the following

- Armistice of Copenhagen of 1547, which ended the Danish war known as the Count's Feud
- Armistice of Stuhmsdorf of 1635 between the Polish-Lithuanian Commonwealth and Sweden
- Peace of Westphalia of 1648 that ended the Thirty Years' War
- World War I

  - Armistice between Russia and the Central Powers, at Brest-Litovsk, 1918 (often called the Treaty of Brest-Litovsk)
  - Armistice with Bulgaria, also known as the Armistice of Solun, September 1918
  - Armistice with Germany (signed at Compiègne), 1918
  - Austrian-Italian Armistice of Villa Giusti ended World War I on the Italian front in early November 1918
  - Armistice of Mudros between the Ottoman Empire and the Allies, 1918

- Armistice of Mudanya between Turkey, Italy, France, Britain, and later Greece, 1922
- World War II

  - Armistice with France (Second Compiègne), 1940
  - Armistice of Saint Jean d'Acre between British forces in the Middle East and Vichy France forces in Syria, 1941

- Armistice with Italy, 1943.
- Moscow Armistice ending the Continuation War, signed by Finland and the Soviet Union on September 19, 1944
- Unconditional surrender implemented by Germany at the end of the war, immediately prior to V-E day on May 8, 1945.
- Japanese Instrument of Surrender on September 2, 1945.

- Armistice Agreements between Israel and its neighbors Egypt, Jordan, Lebanon, and Syria, 1949[24]
- Korean War Armistice, July 1953
- Armistice of Trung Gia ending the First Indochina War, signed by France and the Viet Minh on July 20, 1954
- Armistice in Algeria in an attempt to end the Algerian War, 1962.

Essentially, in the history of U.S. military operations, other than the world wars (which arguably were conditioned on surrenders) and the Korean conflict that never really ended, the use of an armistice to bring war to a termination is uncommon. Of note, from an article from the *New York Tribune* written in 1864 that talks about the folly of an armistice as a way to come to cease hostilities in the Crimean War during the 19th century, see Figure 1.1. This article is a good example of not only the confusion behind what is an armistice, truce, or cease-fire, but also of the fortitude of the individual Russian and French soldiers at the time to bring war to termination.

The oldest term of the three in this category is truce, which in the Middle Ages usually had a religious connotation as in the phrase "Truce of God."[25] Hugo Grotius used *truce* to mean an agreement by which warlike acts are for a time abstained from, though the state of war continues, and if hostilites were resumed after a truce, according to Grotius, there would be no need for a declaration of war, since the state of war was "not dead, but sleeping."[26]

Truces might be concluded by generals in command of forces or by officers of lower rank. In the absence of agreement to the contrary, it was lawful to rebuild walls or to recruit soldiers during a truce, but actual acts of war were forbidden, whether against person or property; that is to say, "whatever is done by force against the enemy."[27] Also forbidden were the bribery of enemy garrisons and the seizure of places held by the enemy. If a truce was violated, the injured party was free to resume hostilities "even without declaring war." Private acts did not constitute a violation, however, unless there was a public command or approval.

When the codification of international law began in the second half of the 19th century, a truce was the procedure by which belligerents entered into parleys, and an armistice was an actual agreement to suspend military operations. According to Articles 43 to 45 of the Brussels Declaration of August 27, 1874, a *parlementaire* was a person who had been authorized by one of the belligerents to enter into communication with the other side. He advanced bearing a white flag, accompanied by a bugler or drummer, sometimes also by a flag bearer. The enemy commander was not in all cases and under all conditions obliged to receive a parlementaire and was, in any case, entitled to take "all measures necessary for preventing the bearer of the flag of truce taking advantage of his stay within the radius of the enemy's position to the prejudice of the latter."[28]

Spontaneous, informal truces have been common during wartime. Enemies came together to listen to music, barter food, play football, and even celebrate Christmas. The Christmas Truce of 1914, which took place a few months after the outbreak of the World War I, across the Western Front, is the most celebrated informal truce of all. But enemies had fraternized before; for example, in the battlefields of Spain during the Napoleonic Wars, in Civil War America, in the valley of Tchernaya during the Crimean War, and in the Second Boer War. The following are examples of truces from history.

Bartering in Spain

The Duke of Wellington, who led the British, Spanish, and Portuguese forces against the French in the Peninsular War of 1807 to 1814, saw with grief the British soldiers fraternizing with the French in many instances. It was not uncommon for sentries or those stationed in outposts to come together for a smoke or a chat and to barter food and other essentials. Groups seen to be foraging in the open were generally not shot at, and there were even exchanges of badly injured prisoners. Amid the many atrocities these informal truces offered a flicker of humanity to the embattled troops.

This is evident in the official dispatches of the Duke of Wellington during this particular campaign. The following is an example taken from the official file and first published in 1834:

> The A.G. [Adjutant General] to Major Gen. Picton, 3d division. 1st March, 1811.[29] "I am to inform you that on the 27th ult.[30] a French officer, with a flag of truce, informed the officer commanding the outposts that an English officer of the 45th regt. had deserted and reached Santarém[31]; but to this report my Lord Wellington attached no credit, till his Lordship heard from me that Lieut. Burke, of the 45th regt., had been absent and unaccounted for since the 23d ult.
>
> Although my Lord Wellington thinks there is no reason to doubt that Lieut. Burke has committed this crime, which hitherto has been unknown among the officers of the British army, it is desirable that, as far as possible, light should be thrown on the possible causes of this conduct, and that Lieut. Burke's actions at and immediately previous to the period of his desertion

should be ascertained and recorded. It is his Lordship's wish therefore that a Court of Inquiry, composed of officers of the 45th regt., should be assembled to investigate this circumstance. They will write down all that occurred, referring to the occasion of Lieut. Burke's quitting the regiment, the hour at which he went, and where he was last seen, where he said he was going, and his alleged motives. The Court will endeavor to ascertain whether any grounds of suspicion were manifested or were entertained lately that he had had correspondence with the enemy. They will learn whether he had any acquaintance in the country or at Lisbon; and if so, who they were; whether he was capable of taking any plans of the works at Alcoentre, or was curious respecting them. In short, it is to be their object to throw light upon this occurrence; and they will for that purpose record on their proceedings whatever information may appear to deserve attention or likely to lead to knowledge of the motives of Lieut. Burke's conduct.[32]

Truces in the Crimean War

In an account of the Crimean War of 1854 to 1856, published in April 1883, the *New York Times* relates instances of friendly communication between French and Russian soldiers. An informal system of white flags of truce was established so that bartering could take place between sentries. It was common for the Russians to leave a bottle of vodka to be collected by the French who left a couple of loaves of white bread in its place. On the appearance of a white flag, firing would cease and the former enemies would raise their heads and exchange smiles, nods, jokes, and goods. A while later, fighting would resume as usual (see Figure 1.1).

Music in Civil War America

Informal truces were widespread in the American Civil War, too. During the War between the States, for instance, Rebels and Yankees traded tobacco, coffee, and newspapers, fished peacefully on opposite sides of a stream, and even gathered blackberries together. Some degree of fellow feeling had always been common among soldiers sent to battle. Often Yankees and Rebels occupying the opposite sides of a river would swim to meet on an island in the middle. They would exchange tobacco and coffee and swap newspapers and would chat a little before resuming shooting. When camping in the same area, Federates and Confederates would even join in concerts and play each other's favorite songs. A shout "Rats in your holes!" was a signal that everyone should take cover again.

Of note is the importance that an infamous truce had on the course of history in defining the beginning of the American Civil War.[33] Fort Pickens was a coastal defense fortification built on Santa Rosa Island at the entrance of to Florida's Pensacola Bay. During the secession crisis of early 1861, U.S. troops refused to surrender the fort to Confederate authorities,

INCIDENTS OF THE CRIMEAN WAR.—The valley of Varnutka and the forest of Baïdar were found to be unoccupied by any Russian detachment except a few Cossacks, who galloped away at sight of the enemy. In the night, however, outposts were placed to watch the French manœuvre, and ascertain its purpose. Capt. Symony, of the Sixth Regiment of Dragoons, being in command of a squadron on vedette duty, perceived a considerable body of Russian cavalry when the moon rose. It was detailing those outposts along the French line. He charged it, and almost entirely cut it to pieces with the sabre. Gen. Pelissier published a general order in praise of Capt. Symony, who also received the cross of the Legion of Honor for his gallant conduct. After this partial encounter the expedition returned to its camp without being attacked by the Russian troops on the Mackenzie Heights, as was expected. I had enjoyed the ride exceedingly, and my immediate chief, as well as the Ambassador, was glad to hear from me all necessary particulars of it. From the hills on each side of the valley of the Tchernaya, which formed a sort of neutral ground after this movement to clear it of detachments, the hostile armies could see each other distinctly without using field-glasses. Communications were soon established between them by signals at the advanced posts. A French sentry would tie his pocket-handkerchief on his bayonet, and a Russian sentry would leave a bottle of *vodka*, or brandy, at the end of his beat. In the evening a comrade not on guard would go to the spot, and, taking the bottle, would put a couple of loaves of white bread in its place. This traffic was carried on with great mutual satisfaction until it became known to commanding officers of regiments, who suppressed it peremptorily. The soldiers were in the habit of seeing white flags raised, and staff officers of the opposed armies meeting amicably to transact whatever business might be on hand, and it was not surprising that they should try to follow the example for the settlement of their own little bargains. After long practice, the management of short armistices under flags of truce was reduced to a perfect system of friendly regularity. A white flag would appear on a rampart or a trench. Firing would cease on both sides. Heads would be raised from parallels and rifle-pits often not many yards distant from each other, and soldiers would see the enemy whom they had been firing at. No ill-feeling was ever shown on either side. They would even smile and nod to each other. The French would cut jokes, which, translated by some Russian officer, would be received by his men with shouts of laughter. Russian infantry soldiers were supplied with better boots for marching in mud and snow than those of the allied armies, who took every opportunity of procuring a good pair from the feet of a dead enemy on the field of battle. This became a jest in the international chats of an armistice. I remember once a diminutive young Russian officer holding up his well-booted foot, and calling out in broken English, with linguistic pride: "Come! Take!" A herculean Captain of the Black Watch replied, with his Highland bonnet in his hand, high in the air: "Accept a fair exchange. You would look better in this!" Much laughing resounded from trench to rifle-pit. When the respective staff officers had finished their conference and saluted each other courteously, they retired to their lines, the flags of truce were lowered, and firing was resumed in the batteries with as much virulence as ever. On one occasion two Sergeants, an Englishman and a Russian, had been making friendly signs to each other, when the truce suddenly ended; the Russian, still standing up inadvertently, would certainly have had at least one rifle-ball through his head if the Englishman had not delayed the fire of his men while he made violent gesticulations to him to go down from the rampart, which he did with a wave of the hand in acknowledgment of his enemy's courtesy.—*Temple Bar.*

Published: April 1, 1883
Copyright © The New York Times

Figure 1.1 *New York Times* Article on Crimean War

and for a time it appeared likely that hostilities might erupt either at Fort Pickens or at Fort Sumter, where similar circumstances prevailed. Ultimately, an informal truce was implemented at Pensacola and the crisis there faded, while Fort Sumter earned notoriety as the location where the Civil War began. Circumstances and the use of a truce clearly shaped history for without it Pickens and not Sumter would have the dubious title as from where the "shot heard around the world" emanated.[34] (See the text box, An Armistice—How It Would Ruin Us, for a *New York Tribune* article on the potentiality of an armistice as a way to end the American Civil War.)

Football in South Africa

It was customarily known that toward the end of Second Boer War (1899 to 1902) in South Africa, a game of football took place between some of the Boers and a British unit under the command of Major Clement Edwards. Sunday truces were common during this war as the Boers for religious reasons abstained from fighting on a Sunday. Most notably a truce on Christmas and Boxing Day was reported to have taken place in Mafeking in 1899.

War was declared on October 11, 1899. The Boers immediately besieged the cities of Ladysmith, Kimberley, and Mafeking—the latter defended by Colonel Robert Baden Powell (who later went on to form the Boy Scouts movement). In December 1899, some 8,000 Boers under the command of Louis Botha defeated 21,000 British troops at the Battle of Colenso. In early 1890, the British suffered more defeats in particular at the Battle of Spion Kop; some years later, supporters of Liverpool Football Club named the popular end of their ground at Anfield "The Spion Kop"—reportedly in honor of the truce offered in the name of humanity that infamous Christmas in the final year of the 19th century.[35]

The Christmas Truce

Arguably the most famous truce of them all is commonly referred to as the Christmas Truce, which started on Christmas Eve of 1914 and lasted until the first days of January 1915. Many letters sent home from the Western Front tell of a "waking dream," a "wonderful day," a Silent Night when fighting stopped and the trenches were illuminated by candles and resounded with Christmas carols sung in different languages. Germans and British ventured in no man's land, exchanged food, and buried their dead. Football matches were said to have been organized as well. Those in charge wanted to make sure that no such truce happened again. However, informal truces continued to take place the following years though on a much smaller scale. On parting, a British soldier said

"AN ARMISTICE—HOW IT WOULD RUIN US"

—*New York Tribune* editorial, September 27, 1864

What is an armistice? Webster defines it to be . . . "a temporary suspension of hostilities *by agreement of the parties.*"

An armistice is the cardinal idea upon which the McClellan movement swings in this Presidential canvass. If McClellan is elected, he will be elected by it.

Suppose he is elected . . . The expectation of an armistice at a future day certain, would as surely break down and dissolve an American army, an army of volunteers fighting for a principle as the flow of the Niagara would dissolve and wash away salt. The 4th of March [the date of the inauguration] would inevitably find us in a condition to do what? To propose an armistice? Oh no! but abjectly, and with just fear and retribution trembling, to receive propositions for a cessation of hostilities. We should be conquered . . .

Who is to take the initiative in . . . the opening of the negotiation, who is to ask for the "convention" and propose the "agreement" [for an armistice]? Not President Davis for he has not asked for an armistice, and he won't ask for an armistice, so long as his heart locks within itself the manhood of courage instead of the sheepishness of cowardice, and so long as his soul remains faithful to the Confederacy which has committed its life to his keeping . . . Who then, is to take the initiative, and send commissioners to propose an armistice? Why President McClellan, clearly.

His commissioners go. They unfold their credentials, *and in the very act of unfolding them recognize the Rebel Confederacy.*

This legal result of the cowardly and traitorous folly of proposing an armistice could not possibly be escaped . . . This fact of recognition by McClellan's administration would immediately be accepted in Paris and London as the solvent of the difficulty which for three years has defeated the application of the Confederate States to be recognized as an independent power. France and Great Britain have consistently replied to [Confederate diplomats John] Slidell and [James] Mason's entreaties: "The American Government treats you as Rebels. Until you can fight yourselves out of it, we can not treat with you as an independent power without getting into war." But the obstacle to this coveted recognition would be removed throughout Europe in an instant by McClellan's proposal of an armistice. France, England, Spain, Austria, and Belgium, would acknowledge the sovereignty of the Confederacy forthwith, and make treaties with them, the commercial classes of which would hourly bribe those powers to help the Rebels while the war lasted.

The argument might well stop here. But let us follow up this negotiation for an armistice. The first question to be settled after the proposal would be Jeff Davis's inquiry . . ., "What is the armistice which you propose?"

"An immediate cessation of hostilities, to the end that peace may be restored on the basis of the Federal Union of the States."

"We will accept the proposal upon the terms and conditions which public law affixes to an armistice. We will withdraw the Confederate troops from every part of your territory; we will suspend the blockade of your coast, and stop privateering on your commerce. You must withdraw the United States troops from every part of our territory; you must suspend the blockade of any and every part of our coast, and cease from capturing merchant vessels bound to our ports. That is, we must be upon terms of equality with you, and free from duress, and relieved from all coercion and restraint, in order to enter into the convention for reunion which you propose."

McClellan's commissioners could not possibly escape from this definition of an armistice, in connection with its object, convention for reunion.

Through the relaxation or suspension of the blockade, and the demoralization of the pickets, supplies of all sorts . . . would get easy ingress and egress into and out of the Confederacy. And in the train of these commissioners would go Delay, stately, cunning, ceremonious, ingenious, diplomatic Delay . . . and the ships of Liverpool, Marseilles, Bremen, and Trieste would the while flock like pigeons to the Southern ports; the cotton, sugar, and tobacco of the Confederacy would get converted into gold; what the Rebellion needed of arms, munitions, clothing, machinery, and men, would be supplied to her; treaties of amity, as well as commerce . . ., would be snug in the State Department at Richmond. The Rebellion, materially re-invigorated and morally braced by the recognition and promised support of the British, French, Spaniards, and Austrians, would be strong enough in September '65, to stalk into the Peace Commission at Richmond in the person of Jeff. Davis, and say: "This affair must come to a conclusion. All negotiations for a peace with the Confederate States must be based upon the recognition of their independence. . . ."

What a condition we would be in? Where would be our army? Desertions consequent on the loss of its spirit, and the destruction of its discipline, sickness and death so sure to run havoc through troops that are idle and demoralized, would have swept it away by whole brigades. Only a decaying skeleton of it would be left. The two hundred thousand black soldiers and employees in the service, would early in March have been kicked out, to appease the beastly rage which shirked in Democratic processions, "This is a white man's war!" The blockade would have to be rescued again by a fleet which had anchored its spirit and vigilance deep down. And when we came to key up the nation to the sacrifice and elasticity necessary to an offensive war could it be done? Every man in this country out of an idiot asylum knows that it could not be done. The war would be gone. The South would triumph.

---

*Source:*http://teachingamericanhistory.org/library/index.asp?document=1478(retrieved December 27, 2009.

to a German: "Today we have peace. Tomorrow you fight for your country. I fight for mine. Good luck."

Stanley Weintraub captured this in his book, *Silent Night: The Story of the World War I Christmas Truce.* He writes that in December 1914 astride the borders of Belgium and France on both sides of the front lines in Flanders, soldiers of two of Queen Victoria's grandsons, Kaiser Wilhelm II and

George V, faced off from rows of trenches that augured a long war of attrition.

For months the fight was intense, but on Christmas day the troops from both sides decided among themselves to not only stop fight but to join hands in camaraderie and song. Opposite, the British applauded each song, and a "big voice" responded from the German parapets, "Blease come mit us into the chorus." After a killjoy on the British side shouted back, "We'd rather die than sing German," the big voice boomed, in English, "It would kill us if you did."[36]

Rightly, the Germans assumed that the other side could not read traditional Gothic lettering, and that few English understood spoken German. "You No Fight, We No Fight" was the most frequently employed German message.[37] Toward midnight, firing ceased and soldiers from both sides met halfway between their positions. "Never," wrote Muhlegg, "was I as keenly aware of the insanity of war."[38] Christmas helped—at least for the moment . . . to bring together men who really, they recognized, didn't hate each other. Their fraternization, dangerously unwarlike from the headquarters' perspective, seemed unstoppable.

Witnesses exclaimed, What a sight!—little groups of Germans and British extending almost the length of our front! Out of the darkness we could hear laughter and see lighted matches. . . . Where they couldn't talk the language, they were making themselves understood by signs, and everyone seemed to be getting on nicely. Here we were laughing and chatting to men whom only a few hours before we were trying to kill![39] According to his diary, [Field Marshal Sir John French] "issued immediate orders to prevent any recurrence of such conduct, and called the local commanders to strict account," which resulted, he understated later, "in a good deal of trouble."[40]

A London Rifles entrepreneur, offering large quantities of appropriated bully beef and jam, acquired a prized *Pickelhaube* [spiked helmet], which was nearly impossible to conceal or cart home. The day after Christmas he heard someone shouting for him from the German side. Whom he eventually met in No Mans Land and had the following conversation, "Yesterday," his new friend appealed, "I give my hat for the Bullybif. I have grand inspection tomorrow. You lend me and I bring it back after." Somehow the deal was kept.[41]

The Germans came out of their protective holes, fetched a football, and invited our boys out for a little game. Our boys joined them and together they quickly had great fun, till they had to return to their posts. I cannot guarantee it, but it was told to me that our lieutenant colonel threatened our soldiers with machine guns. Had just one of these big mouths gathered together ten thousand footballs, what a happy solution that would have been, without bloodshed.[42]

Perhaps most importantly was that many troops discovered through the truce that the enemy, despite the best efforts of propagandists, were not monsters. Each side had encountered men much like themselves, drawn from the same walks of life—and led by professionals who saw the world through different lenses. A future general, Captain Jack of the Cameronians, averse to the truce when on the line, had speculated in his diary a few days earlier, about the larger implications of the cease-fire, which had extended farther than governments conceded. "It is interesting to visualize the close of a campaign owing to the opposing armies—neither of them defeated—having become too friendly to continue the fight."[43]

During a House of Commons debate on March 31, 1930, Sir H. Kingsley Wood, a Cabinet minister during the next war, and major "in the front trenches" at Christmas 1914, recalled that he "took part in what was well known at the time as a truce. We went over in front of the trenches, and shook hands with many of our German enemies. A great number of people [now] think we did something that was degrading."[44]

Refusing to presume that, he went on, "The fact is that we did it, and I then came to the conclusion that I have held very firmly ever since, that if we had been left to ourselves there would never have been another shot fired. For a fortnight, the truce went on. We were on the friendliest terms, and it was only the fact that we were being controlled by others that made it necessary for us to start trying to shoot one another again." [Quoting from a ballad] "The ones who call the shots won't be among the dead and lame, and on each end of the rifle we're the same."[45]

The friendly meshing of troops across the front occurs at some moment in most wars. An Australian trooper remembered just this in his tenure in World War II in North Africa late in 1941. At Tobruk . . . for a few hours, men on both sides openly stood up, in the past an invitation to be shot. They dried their clothes, made tea, and did not return desultory fire. An infantryman looking unnaturally old after months of exposure to the cruel sun, the dust, and the strain, looked across at the enemy, who were suffering in the same way, and said to no one in particular, "Nobody said we couldn't like them, they just said we had to kill them. All a bit stupid, isn't it?"[46]

FORMAL PEACE TREATIES

Peace treaties are agreements between two hostile parties, usually countries or governments that formally end an armed conflict. It is different from an armistice, which is an agreement to take steps to demilitarize the situation. There are many possible issues that may be included in a peace treaty, and a treaty's content usually depends on the nature of the conflict being concluded. Some of these may be:

- Formal designation of borders
- Processes for resolving future disputes
- Access to and apportioning of resources

- Status of refugees
- Settling of existing debts
- Defining of proscribed behavior
- The re-application of existing treaties.

Treaties are often ratified in territories deemed neutral in the previous conflict and delegates from these neutral territories act as witnesses to the signatories. In the case of large conflicts among numerous parties, there may be one international treaty covering all issues or separate treaties signed between each party.

In modern times, certain intractable conflict situations may first be brought to a cease-fire and are then dealt with via a peace process where a number of discrete steps are taken on each side to eventually reach the mutually desired goal of peace and the signing of a treaty. A peace treaty also is often not used to end a civil war, especially in cases of a failed secession, as it implies mutual recognition of statehood. In cases such as the American Civil War, it usually ends when the armies of the losing side surrender and the government collapses. By contrast, a successful secession or declaration of independence is often formalized by means of a peace treaty.

Peace treaties in the modern era have become less and less in vogue mainly because of the time requirements demanded to negotiate between parties. For instance, the 1978 Camp David Peace Accords took roughly 12 days to come to fruition once both Israel's Menachem Begin and Egypt's Anwar el Sadat took to the leadership of U.S. President Jimmy Carter.

Although used less and less in the determination of strategy to terminate war, the treaties that have existed based on conflict in the past have mostly had a significant impact on the future of war and its operating environment. For instance, one of the many explanations historians have given is that the origins of World War II are directly attributable to the Treaty of Versailles—that Versailles was much too harsh a peace, and that the Germans should have been given an easy peace to bring them into the European community.[47] The Treaty of Versailles radically altered the geography of Europe. As a result the Germans felt betrayed for the Treaty clearly laid the blame on German; thus, it was accordingly punished. Hitler later capitalized on this by vowing to make Germany great again.

Some would argue that it wasn't the direct cause; some would say it was one of the most important factors. Regardless, the restrictions imposed by the Versailles Treaty created a social and economic depression in Germany, with widespread disillusion among its people. These are the conditions in which radical and extremist ideas take root and become popular. The population was looking for (indeed desperate) for change; unfortunately, it happened to be the Nazis and Adolf Hitler that came along and promised the change and reform that they were looking for and told them what they wanted to hear. When the Nazis started building new weapons in secret in the mid-1930s, they put Germany's workforce back to work. Suddenly there were jobs and apparent prosperity and the German people could see a brighter future. This helped sell the belief that the Nazis and their policies were righteous and that Hitler was a great leader.

The rest is history. The terms of the treaty made another war almost inevitable. Several terms were arguably blatantly shortsighted. First, the forced signature required of Germany placing full blame for the war on her shoulders. Second, the faulty border drawing by the imperialistically motivated powers of France and Britain created false nations such as Czechoslovakia, Yugoslavia, Kuwait, Saudi Arabia, and others and vast land grabs by both nations as they took over various colonies of Germany plus splitting up the Middle East, creating problems that still exist today. Third, the realignment of borders left large German minorities under the rule of other ethnic groups. Finally, the attempted destruction of German military power allowed opportunistic, expansionist countries like Poland to take advantage of its weak neighbor.

Ancient History

One of the earliest recorded peace treaties was concluded between the Hittite and Egyptian empires after the ca.1274 BCE Battle of Kadesh. The battle took place in what is modern-day Syria, the entire Levant being at that time contested between the two empires. After an extremely costly four-day battle in which neither side gained a substantial advantage, both sides claimed victory. The lack of resolution led to further conflict between Egypt and the Hittites with Ramesses II capturing the city of Kadesh and Amurru in Year 8 of his reign.[48]

However, the prospect of further protracted conflict between the two states eventually persuaded both their rulers, Hatusiliš III and Ramesses to end their dispute and sign a peace treaty. Both sides could not afford the possibility of a longer conflict since they were threatened by other enemies. Egypt was faced with the task of defending her long western border with Libya against the incursion of Libyan tribesmen by building a chain of fortresses stretching from Mersa Matruh to Rakotis while the Hittites faced a more formidable threat in the form of the Assyrian Empire which "had conquered Hanigalbat, the heartland of Mitanni, between the Tigris and the Euphrates" rivers that had previously been a Hittite vassal state.[49]

The peace treaty was recorded in two versions, one in Egyptian hieroglyphs, the other in Akkadian, using cuneiform script; fortunately, both versions survive. Such dual-language recording is common to many subsequent treaties. This treaty differs from others, however, in that the two language versions are differently worded. Although the majority of the text is identical, the Hittite version claims that the Egyptians came suing for peace, while the Egyptian version claims the reverse. The treaty was given to the Egyptians in the form of a silver plaque, which was taken back to Egypt and carved into the Temple of Karnak.

The treaty was concluded between Ramesses II and Hatusiliš III in Year 21 of Ramesses' reign (ca.1258 BCE).[50] Its 18 articles call for peace between Egypt and Hatti and then proceeds to maintain that their respective gods also demand peace. It contains many elements found in more modern treaties, although it is perhaps more far-reaching than later treaties' simple declaration of the end of hostilities. It also contains a mutual-assistance pact in the event that one of the empires should be attacked by a third party or in the event of internal strife. Additionally, there are articles pertaining to the forced repatriation of refugees and provisions that they should not be harmed; this might be thought of as the first extradition treaty. There are also threats of retribution, should the treaty be broken. This treaty is considered of such importance in the field of international relations that a reproduction of it hangs today in the UN headquarters.

Modern History

Other famous examples of treaties include the 1815 Treaty of Paris signed after Napoleon's defeat at the Battle of Waterloo, and the Treaty of Versailles, formally ending World War I. The latter treaty is possibly the most notorious of peace treaties, in that it is "blamed" by some historians for the rise of National Socialism in Germany and the eventual outbreak of World War II. The costly reparations that Germany was forced to pay the victors, the fact that Germany had to accept sole responsibility for starting the war, and the harsh restrictions on German rearmament were all listed in the treaty and caused massive resentment in Germany. Whether the Treaty of Versailles can be blamed for starting another war or not, it shows the difficulties involved in making peace.

Another famous example would be the series of peace treaties commonly linked by the fact that they brought the Thirty Years War to an end, is known as the Peace of Westphalia. It is often said that the Peace of Westphalia initiated the modern fashion of diplomacy as it marked the beginning of the modern system of nation-states. Today it is viewed as responsible for hegemonic ambitions of individual states that emerged from the treaty. Subsequent wars were no longer over religion, but rather revolved around issues of state. This allowed Catholic and Protestant powers to ally, leading to a number of major realignments.

JOINT POLITICAL AGREEMENTS

We have already discussed two categories of the six for our war termination strategies. The third mechanism defined by Clarke and Staudenmaier is the joint political agreements. This type of agreement between parties usually stipulates how the conflict will end and may indicate how "peace" will be maintained. According to Clarke, one clear example of such a political agreement is the "Agreement on Ending the War and Restoring Peace in Viet-Nam" (see Appendix A) that provides for a cease-fire throughout Vietnam, withdrawal of U.S. troops, release of prisoners of war, restoration of the demarcation line between North and South Vietnam, and the creation of an international body to supervise the truce.[51]

This type of war termination activity allows each side to extend itself politically (meeting the third leg of the Clausewitzian trinity) to agree on a solution to the ongoing fighting. Note again that this typically is not preplanned or forecasted as a potential outcome.

In the waning years of the Vietnam War in September 1970, Secretary of State Henry Kissinger agreed that Northern troops could remain in the South after a settlement. Such a "cease-fire in place" would allow the communist forces to renew the offensive once the Americans left. After a major offensive in the spring of 1972 by the North against the South (the Easter Offensive), negotiations began to move forward and by mid-October, Kissinger and his North Vietnamese counterpart worked out a draft joint political agreement that called for the remaining U.S. Forces to depart Vietnam and the return of U.S. prisoners of war while deferring ultimate decisions about the South's political future.[52]

In the case of the Vietnam War, the United States, South Vietnam, the Viet Cong, and North Vietnam formally signed "An Agreement Ending the War and Restoring Peace in Vietnam" in Paris on January 17, 1973. Due to South Vietnam's unwillingness to recognize the Viet Cong's Provisional Revolutionary Government, all references to it were confined to a two-party version of the document signed by North Vietnam and the United States; the South Vietnamese were presented with a separate document that did not make reference to the Viet Cong government. This was part of Saigon's long-time refusal to recognize the Viet Cong as a legitimate participant in the discussions to end the war.

As with the definition of this category, an agreement to stop fighting was politically agreed on in a joint fashion by both sides. The settlement included a cease-fire throughout Vietnam. It addition, the United States agreed to the withdrawal of all U.S. troops and advisers (totaling at the time about 23,700) and the dismantling of all U.S. bases within 60 days. In return, the North Vietnamese agreed to release all U.S. and other prisoners of war.

Both sides agreed to the withdrawal of all foreign troops from Laos and Cambodia and the prohibition of bases in and troop movements through these countries. It was agreed that the demilitarized zone (DMZ) at the 17th Parallel would remain a provisional dividing line, with eventual reunification of the country "through peaceful means." An international control commission would be established made up of Canadians, Hungarians,

Poles, and Indonesians, with 1,160 inspectors to supervise the agreement. According to the agreement, South Vietnamese President Nguyen Van Thieu would continue in office pending elections. Agreeing to the South Vietnamese People's right to self-determination, the North Vietnamese said they would not initiate military movement across the DMZ and that there would be no use of force to reunify the country.[53]

In another example, Kosovo Peace Accords followed by the U.S.-led air raid (March 24, 1999) on the Federal Republic of Yugoslavia (Former Yugoslav Republic or FYR, Serbia, and Montenegro), including Kosovo, which the North Atlantic Treaty Organization (NATO) regarded as a province of Serbia. On June 3, NATO and Serbia reached a Peace Accord after the United States declared victory, having successfully concluded its 10-week struggle to compel Slobodan Milosevic to capitulate.[54]

While declaring victory, Washington did not yet declare peace: the bombing continued until the victors determined that their interpretation of the Kosovo Accord had been imposed. The Rambouillet Agreement is the name of a proposed peace agreement between then-Yugoslavia and a delegation representing the ethnic-Albanian majority population of Kosovo. It was drafted by NATO and named for Chateau Rambouillet, where it was initially proposed. The significance of the agreement lies in the fact that Yugoslavia refused to accept it, which NATO used as justification to start the Kosovo War. Belgrade's rejection was based on the argument that it contained provisions for Kosovo's autonomy that went further than the Serbian/Yugoslav government saw as reasonable.

The most controversial provision was that Kosovo would only de jure be a province of Serbia and de facto a third republic, but with greater autonomy than Serbia and Montenegro with the respect to their relations to the Serbian/Yugoslav federal government. Serbia viewed this as secession of Kosovo from Serbia, but the most controversial aspect of this was that while Serbia would have no influence over its southern province whatsoever, Kosovo would have substantial influence over Serbia. Kosovars would take part in Serbian elections and its members of parliament (MP) would sit in a parliament that would have no jurisdiction over Kosovo. Kosovars would have guaranteed seats in the Serbian Government and the Serbian Supreme Court, which would deal exclusively to the territory of Serbia without Kosovo. Kosovo would also have an independent judicial system including its own autonomous Constitutional Court, but it would also have guaranteed representatives in the Yugoslavian judiciary, which would have no jurisdiction over Kosovo. Furthermore, NATO would have completely free and unrestricted military access to the entire country very closely approaching a level that could be considered to amount to an occupation.

The Serbian Parliament responded on March 23, 1999, to the agreement with a sharp criticism. Though it agreed that Kosovo should be given autonomy, it stated that it would prefer the incursion of the UN over that of NATO.

The Rambouillet Agreement was important for the debate about the Kosovo War. Although it was initially fully rejected by the Kosovar Albanian side, Belgrade had agreed to all of the political and nonmilitary points. Belgrade requested NATO troops be replaced with UN troops for full acceptance. At the talks, the Agreement was repeatedly amended until the Kosovo Albanian side was forced to sign, whereas Belgrade rejected it. The biggest problem for both sides was that the Contact Group's non-negotiable principles were mutually unacceptable. The Albanians were unwilling to accept a solution that would retain Kosovo as part of Serbia. The Serbs did not want to see the pre-1990 status quo restored, and were implacably opposed to any international role in the governance of the province. The negotiations thus became a somewhat cynical game of musical chairs, each side trying to avoid being blamed for the breakdown of the talks.

In the end, on March 18, 1999, the Albanian, American, and British delegations signed what became known as the Rambouillet Accords, while the Serbian and Russian delegations refused. The Accords called for NATO administration of Kosovo as an autonomous province within Yugoslavia; a force of 30,000 NATO troops to maintain order in Kosovo; an unhindered right of passage for NATO troops on Yugoslav territory, including Kosovo; and immunity for NATO and its agents to Yugoslav law.

Critics of the Kosovo war have claimed that the Serbian refusal was prompted by unacceptably broad terms in the access rights proposed for the NATO peacekeeping force. This was based on standard UN peacekeeping agreements such as that in force in Bosnia, but would have given broader rights of access than were really needed—onto the entire territory of Yugoslavia, not just the province. It has been claimed that Appendix B would have authorized what would amount to a NATO occupation of the whole of Yugoslavia, and that its presence in the Accords was the cause of the breakdown of the talks. The chapter dealing with the Kosovan economy was also equally revealing. It called for "privatization of all Government assets"; this seems to be commensurate with the fact that around 372 centers of industries were bombed during the conflict, including many with no relevance to military means.

Events proceeded rapidly after the failure at Rambouillet. The international monitors from the Organization for Security and Cooperation in Europe withdrew on March 22, 1999, for fear of the monitors' safety ahead of the anticipated bombing by NATO. On March 23, the Serbian assembly accepted the principle of autonomy for Kosovo and the nonmilitary part of the agreement. But the Serbian side had objections to the military part of the Rambouillet agreement, Appendix B in particular, which it characterized as "NATO occupation." The full document was described "fraudulent" because the military part of the agreement was offered only at the very end of the talks without much possibility for negotiation, and because the other side, condemned in harshest terms as a "separatist–terrorist delegation," completely refused to meet the delegation of FRY and negotiate directly during the Rambouillet talks at all. The following day, March 24, NATO bombing began.[55]

DECLARATION OF A UNILATERAL VICTORY BY VICTOR

The fourth category defines a situation where the fight is truly unbalanced. When the very existence of the state is at stake, the conflict can end by the victor unilaterally declaring victory. While Clarke uses Vietnam as a potential example, it is clear that Operation Desert Storm stands out as a prime example for this category of war termination.[56] Once the U.S.-led coalition expelled Saddam Hussein's forces from Kuwait after a 100-hour ground war, the United States, led by General Norman Schwarzkopf, was drawn to a decisive conclusion by President George H. W. Bush.

The invasion of Kuwait by Iraqi troops that began August 2, 1990, was met with international condemnation, and brought immediate economic sanctions against Iraq by members of the UN Security Council. President Bush deployed American forces to Saudi Arabia and urged other countries to send their own forces to the scene. An array of nations joined the Coalition of the Gulf War. The great majority of the military forces in the coalition were from the United States, with Saudi Arabia, the United Kingdom, and Egypt as leading contributors, in that order. Around $40 billion of the $60 billion cost was paid by Saudi Arabia.

The initial conflict to expel Iraqi troops from Kuwait began with an aerial bombardment on January 17, 1991. This was followed by a ground assault on February 23. It was a decisive victory for the Coalition Forces that liberated Kuwait and advanced into Iraqi territory. The coalition ceased their advance and declared a cease-fire 100 hours after the ground campaign started. Aerial and ground combat were confined to Iraq, Kuwait, and areas on the border of Saudi Arabia. However, Iraq launched missiles against coalition military targets in Saudi Arabia. Then again, on May 1, 2003, some 12 years later, aboard the USS *Abraham Lincoln*, President George W. Bush declared that "major combat operations in Iraq have ended."[57]

Much analysis has been done on tethering the termination of war to the causes of conflict. One study suggests that "examining war termination in the Persian Gulf through the prisms of interests, fear, and honor (which Thucydides identified 2,400 years ago as what he considered the three causes of war), the Gulf War and the high intensity conflict in the first few months of Operation Iraqi Freedom were overwhelmingly one-sided events and clear coalition victories that the U.S.-led coalition could reach a unilateral conclusion that they had "won." The premise for this theory is that the greater the perceived leverage in these three areas, the more "satisfactory" the resulting peace.[58]

During the First Gulf War, President Bush defined the goals of the Coalition to cause ". . . the immediate and unconditional withdrawal of all Iraqi Forces from Kuwait. Second, the restoration of Kuwait's legitimate government. Third, security and stability for the Gulf." Fourth was the protection of American citizens abroad. In the end, based on pure military terms, the Gulf War was overwhelmingly a one-sided event and a clear coalition victory.[59]

CAPITULATION BY THE LOSER

The fifth form of conflict termination is capitulation. World War II may have ended in a capitulation that was then followed by a series of treaties. Clarke explains that a capitulation may also be accompanied by a declaration of victory or some other termination agreement.[60] Regardless, in this form of termination the "losing" side basically declares defeat.

WITHDRAWAL OF ONE OF THE PARTIES

The final form of conflict end is when one party simply and unilaterally withdraws from active participation in the conflict. Somalia's withdrawal from Ethiopia in 1978 and then the U.S. withdrawal from Somalia after the 1993 Black Hawk Down debacle are prime examples.

Although these six candidates are worthy of consideration for compartmentalizing war termination strategies, Clarke does not go beyond the cessation of hostilities in defining war termination. What Clarke defines are really goals for the military operations and not for the "whole of government" concept required to ending a conflict or bringing it to a satisfactory result.

CONFLICT RESOLUTION

My argument is that these "endings" should include what others would term *conflict resolution* categories, or in other words, what happens to the war-torn country after the fight? Much like Clarke presented six categories for terminating a war, I present the categories below that define the second part of my premise: conflict resolution.

1. **Nation-Building:** This conflict resolution category refers to the process of constructing or structuring a national identity using the power of the state. This process aims at the unification of the people or peoples within the state so that it remains politically stable and viable in the long run. Nation-building can

involve the use of propaganda or major infrastructure development to foster social harmony and economic growth.

Countries going from chaos to order need much more time than it takes for the ink to dry on a pact. Getting a country up and running requires years and many resources, including large sums of money from the international community. The basic structure of a country is the same: political, economic, taxation and judicial systems; infrastructure; cultural, educational, and medical institutions; and more. Because these are so interconnected, fitting them together into a unified, organic whole is a complex undertaking.[61]

2. **Stability, Security, Transition, and Reconstruction Operations:** This is a core U.S. military mission that includes activities across the peace–war spectrum that are conducted to establish or maintain order in states or regions to achieve sustainable peace, while advancing U.S. interests. When transitioning to stability, security, transition and reconstruction operations, a nation focuses on achieving four endstates through political shaping operations, constabulary-type security operations, and state-building.

The first is security for the host-nation population at the local level, produced through a combination of foreign and indigenous forces. The second is political stability, which is a function of creating legitimacy for the new political order and an effective process for inclusion and collective decision making for the society. Stabilization pertains to promoting activities that ready a situation or prepare the ground for a longer-term agenda. This process lays the tracks and prevents backsliding or eruption to greater conflict. The third is reconstruction, both of state institutions and a framework and system for wealth generation. The fourth is the rule of law, creating institutions to provide impartial enforcement of the law and conflict resolution of private disputes, strengthening systems to ensure integrity of personnel within state institutions, and propagating a positive ethos and culture of lawfulness, in collaboration with multiple local indigenous sectors. In pursuing these end states, it is vital to recognize that adversarial political players—both indigenous and regional—will be competing to disrupt these efforts and to achieve alternative end states.[62]

3. **Economic Development:** Like the great British strategist J.F.C. Fuller wrote:

> The first duty of the grand strategist is . . . to appreciate the commercial and financial position of his country; to discover what its resources and liabilities are. Secondly, he must understand the moral characteristics of his countrymen, their history, peculiarities; social customs and systems of government, for all these quantities and qualities form the pillars of the military arch which it is his duty to construct.[63]

Economic development is the sustained increase in the economic standard of living of a country's population, normally accomplished by increasing its stocks of physical and human capital and improving its technology.

4. **Humanitarian Relief:** Humanitarian aid is material or logistical assistance provided for the promotion of human welfare and the advancement of social reforms, typically in response to humanitarian (disease, hurricanes, floods,

earthquakes, etc.) crises. The primary objective of humanitarian aid is to save lives, alleviate suffering, and maintain human dignity. It may therefore be distinguished from development aid, which seeks to address the underlying socioeconomic factors that may have led to a crisis or emergency.

Humanitarian relief and economic development are oftentimes confused to instill the same endstate. Many nongovernmental organizations will use humanitarian relief efforts to uncover the underpinnings to their genesis—that being a poor economy. What oftentimes happens then is the "mission creep" associated with trying to make the humanitarian mission into a larger-scoped operation.

5. **Transitioning Security to the Indigenous Force:** The effort is to establish a local security and national security force that can first partner with foreign assistance or occupation forces and then take the lead in the security of its own nation. This endeavor is critical; for other efforts to proceed unhindered by potential residual violence, there is a great need to establish security in the region. Although an occupation force can easily be trapped in a quagmire providing such security, history has shown that the sooner it can transition to a capable indigenous security force the better for the conflict resolution phase of the operation.

6. **Establishing a Democratic Nation:** For almost a century in the West, democracy has meant liberal democracy. Liberal democracies are political systems marked not only by free and fair elections, but also by the rule of law, a separation of powers, and the protection of basic liberties of speech, assembly, religion, and property. As the strategic-minded author Fareed Zakaria writes, ". . . this later bundle of freedoms—what might be termed constitutional liberalism—is theoretically different and historically distinct from democracy."[64] The shifting from a tyrannically led dictatorship, a socialist government, or even a monarch to democracy is not an easy task. To define it as a conflict resolution endstate needs much planning and resource integration early in the conflict.

7. **Countering an Insurgency and Maintaining a Lasting Presence:** Anyone who has fought an insurgency will tell you that to prevent an insurgency is the first order of business when attempting to resolve and terminate a conflict. History is replete with examples where high-intensity conflict at some point (and oftentimes by surprise) develops into an insurgency where the insurgents are difficult to identify and the tactical, operational, and strategic plans are difficult to define and execute. Once an insurgency gains strength, the ability to terminate the fight and bring resolution to a war-torn nation becomes exponentially more problematic.

8. **Invading a Country to Establish an Imperial Footprint or Continual Military Presence:** In other words, to conquer a nation and deny the indigenous involvement in ruling its own people. This type of ruthless behavior surely results in the population attempting to repulse the conquerors, thus leading to an insurgency. The difference between this category and the one prior is the insurgency in this case takes on the role as freedom fighters, attempting to fight for what was rightly and internationally recognized as theirs.

9. **Providing Persistent Foreign Internal Defense:** Foreign internal defense (FID) is the participation by civilian and military agencies of a government in

any of the action programs taken by another government or other designated organization, to free and protect its society from subversion, lawlessness, and insurgency. Commensurate with U.S. policy goals, the focus of all U.S. FID efforts is to support the host-nation's program of internal defense and development.

These national programs are designed to free and protect a nation from subversion, lawlessness, and insurgency by emphasizing the building of viable institutions that respond to the needs of society. The most significant manifestation of these needs is likely to be economic, social, informational, or political; therefore, these needs should prescribe the principal focus of U.S. efforts. The United States will generally employ a mix of diplomatic, economic, informational, and military instruments of national power in support of these objectives. Military assistance is often necessary to provide the secure environment for the above efforts to become effective.[65] In essence, this category takes on the characteristics of a number of other categories (economic development, reconstruction, countering an insurgency, etc.) and rolls them into a single conflict resolution construct. The timing of an occupying force moving toward FID from, say, partnership with indigenous security forces has been the challenging, albeit, critical decision in recent forays.

10. **Withdrawing under Fire without a Deliberate Transition Plan:** This category can be renamed "cutting and running" for that is what it clearly means. This type of conflict resolution is mostly never planned for but is a byproduct of a failed incursion. One of the most infamous examples in modern times is the U.S. departure from Somalia in 1993 after a staunch warlord stance caused 18 U.S. casualties in October of 1993, resulting in a loss of U.S. public support and a U.S. President unwilling to pursue the nation's original objectives. Therefore, he ordered the withdrawal of all U.S. Forces, essentially causing a failed nation state to occur in eastern Africa. Although this is arguably also a war termination strategy as described in the previous section, the addition of "do nothing" to assist in resolving the woes of the nation affected by the conflict after a hasty withdrawal of troops make this a conflict resolution category.

THE FOG OF TODAY'S ONGOING WARS

Using the current conflicts in Iraq and Afghanistan as a backdrop, an important part of this chapter is the analysis of some of the *excuses* made as to why devising precise exit strategies for Operations Iraqi Freedom and Enduring Freedom were so important prior to their beginning. Although both wars were singled out as tactical successes in the initial battles, the U.S.-led Coalition fumbled through the nesting of goals and objectives for each theater as they related to the national security strategies (NSS) and military strategies in failing to plan for terminating conflict and coming to conflict resolution. The fog of war was clearly on the planners for what was once termed the *Global War on Terror*.

An examination of the Coalition Provisional Authority (CPA) in the post-Paul Gardner era in Iraq, and the Combined Joint Task Force assembled in the early stages in Afghanistan is instructive in having a sense of how the fog of *postwar* had an impact on the fight. For instance, the Iraqi minister of finance and senior adviser to the Prime Minister, Ali A. Allawi, who in 2004 was named the first postwar minister of defense, writes in a recently published book that "more perceptive people knew instinctively that the invasion of Iraq would open up the great fissures in Iraqi society." According to Allawi, what followed was the "rank amateurism and swaggering arrogance"[66] of the occupation, under L. Paul Bremer's CPA, which took big steps with little consultation with any Iraqis—steps Allawi and many others see as blunders. Among Allawi's criticisms are the following:

- The Americans disbanded Iraq's army, which Allawi said could have helped quell a rising insurgency in 2003. Instead, when the army was disbanded, hundreds of thousands of angry demobilized men became a recruiting pool for the resistance. We also note that though the coalition was quick to disband the army, it was reluctant to outlaw, disband, or even marginalize, in some cases, the many militia elements sprouting up in the area of responsibility.

- Purging tens of thousands of members of toppled President Saddam Hussein's Baath party from government, school faculties, and elsewhere left Iraq short on experienced hands at a crucial time. It also left those disenfranchised citizens, albeit former Saddam sympathizers (if one believes being in the Baath party is a litmus test for being a Saddam supporter), without hope for their future. The insurgency, especially the Al Qaeda elements in Iraq, fed on the deep resentment Sunni Arabs felt at their loss of power and prestige. Their feeling of hopelessness has been aggravated by the fact that the force that overthrew Saddam seemingly achieved the impossible—the dethronement of the community from centuries of power—in favor of, as they saw it, an unruly mob led by pro-Iranian clerics. The Sunni Arabs' refusal to tolerate any serious engagement with the new political order has effectively pushed them into a corner and played into the hands of their most determined enemies. It has taken more than four years for the coalition to realize that a reconciliation plan is necessary to reduce the violence associated with the sectarian conflict.

- An order consolidating decentralized bank accounts at the Finance Ministry bogged down operations of Iraq's many state-owned enterprises.

- The CPA's focus on private enterprise allowed the "commercial gangs" of Saddam's day to monopolize business.

- The new government's free-trade policy allowed looted Iraqi capital equipment to be smuggled across international borders.

- The CPA perpetuated Saddam's fuel subsidies, selling gasoline at giveaway prices and draining the budget.[67]

Although the failure of the CPA was pivotal in the early part of the war, what the U.S.-led Coalition didn't grasp early on in Iraq was that the enemy was becoming more lethal and more creative as well as adaptive. Instead, the Coalition focused more on the establishment of rules and the development of a garrison, developing an occupation type of mindset.

One of the biggest mistakes early on in Iraq was that the Coalition didn't immediately recognize this shift in enemy tactics, and it underestimated and perhaps underprepared for the potential of the evolution of an insurgency. Only over time has it come to a slow and unfocused realization of how the threat in Iraq had evolved. The failure to see the emergence of an insurgency set the Coalition back a number of years in coming to a postexecution conflict resolution plan.

In practical terms, in mid- to late 2003, the invading Coalition Forces, which were introduced to destroy a standing army, needed to identify the change in enemy tactics and then transition their actions from a high-intensity fight to that of defeating an insurgency. Instead of switching to an unconventional approach to defeat the insurgency, however, the Coalition maintained a conventional style in most of its engagements, all the while building bureaucratic systems to emulate garrison activities found on installations in the United States and other military compounds throughout the world. Meanwhile, what was supposed to be a straightforward process of overthrowing a dictatorship and replacing it with a liberal-leaning, secular democracy under the benign tutelage of the United States instead turned into an existential battle for identity, power, and legitimacy that was and is affecting not only Iraq, but the entire volatile state system in the Middle East. The CPA's failures and the "allowance" of an insurgency to thrive were arguably the biggest failures in the early parts of the Iraqi war.

The early failures in Afghanistan may be simpler. In short, since 2001 the forces fighting in Operation Enduring Freedom were never completely resourced to accomplish the clear-hold-build construct of defeating an insurgency. For years, the focus was just on killing the bad guys. Afghanistan's terrain and size are formidable. Critics of the war thus far have complained that the size of the force and contiguous placement of units has been woefully short when it comes to troop-to-task (or how many soldiers are required to accomplish all tasks) requirements. Not until 2009 and 2010 when the surge was requested, authorized, and supported were there adequate resources to win the fight in Operation Enduring Freedom. Clearly a lack of proper planning early on in the fight led to this shortfall. General Stan McChrystal was introduced to establish a counterinsurgency strategy. He did that but then was replaced by General David Petraeus in midstream. Petraeus is clearly capable of devising a plan, a la Iraq, to achieve victory but the jury remains out on the support he will get from the whole-of-nation approach that is needed in the end of a decade of fighting in that country.

As Anthony H. Cordesman from the Center of Strategic and International Studies wrote in September of 2009,

> The U.S. failed to properly address many of the most critical grand strategic
> issues it faced in going to war in both Afghanistan and Iraq, the potential

costs and risks involved, the need to address the chance for insurgency and civil conflict, and both the requirement for armed nation-building and for far more careful calculation of whether a major U.S. military intervention was the proper solution to the national security problems involved.

The problem with strategists like Cordesman, however, is that his analysis quickly delves into whether the United States should have gone to war or not. Regardless, he has many insights worth exploring.

For instance, he believes that the United States went to war in both Afghanistan and Iraq seeking to avoid nation-building, and was unprepared for both nation-building and counterinsurgency. It failed to assess the problems in trying to change foreign cultures, governments, economies, and security structures. If it had, it would have certainly looked toward the aftermath of the fight. It then failed to understand the nature of the insurgencies and levels of conflict it faced, the complexity of the actions needed to succeed, and the resources required. It mirror-imaged values and goals that Afghans and Iraqis did not broadly share, and never properly assessed its own ability to staff, resource, and manage the actions it did take—particularly at the civil level.

More of Cordesman's insightful discoveries include that the United States confused holding elections and creating new formal structures of central government with actual effective governance and political accommodation and stability. Its approach to instant democracy and unrealistic approaches to the rule of law and development rather than meeting popular needs laid much of the groundwork for failure in both countries and helped to empower both insurgencies. It is clear that the United States failed to exploit the "golden hour" in both Afghanistan and Iraq when the insurgency was still beginning and when decisive action might well have halted it or sharply limited its impact. The insurgency is something that may have been prevented if the appropriate planning and subsequent war gaming took place prior to the war.

The military, however, proved to be far more effective and adaptable than the civilians. Even today, the presence of some extraordinary civilians in the U.S. embassies and aid efforts in Afghanistan and Iraq cannot disguise the fact that the State Department as an institution is still unable to plan and execute an effective civil effort in both countries. In Cordesman's mind, the United States solved its alliance problems in Iraq by having its allies leave, albeit at what may be a lasting cost to both their future willingness to support the United States and governments that will directly support the United States in future wars. He goes on to surmise that in the case of Afghanistan, NATO/International Security Assistance Force (ISAF) is close to an alliance of the impossible. It divides the military effort into national efforts by committed, standby, and peripheral forces— each of which adopts a different approach to both security and working with civilian aid workers like the Provincial Reconstruction Teams that are critical to the hold and build phases of U.S. and NATO/ISAF strategy.[68]

But the hard power requirements in the form of troops, equipment, and other forms of combat power are only part of the deficit. As discussed, hard power, soft power, and smart power all seem to be hot topics with policy thinkers recently. With Secretary of State Hillary Clinton advocating the employment of soft power to accompany the might of the military, we may see an organization established, and more importantly resourced, to provide the skill sets and talents necessary to conduct postconflict resolution in the form of nation-building and reconstruction.

The recently formed State Deparment's Coordination, Reconstruction, and Stabilization (S/CRS) service originally developed in concept in 2005 by Bush's National Security Presidential Directive (NSPD)-44 took three years to fund, getting an allocation of funds finally in the summer of 2008. In Chapter 5, there is a detailed discussion on the Project on National Security Reform (PNSR); an important element to understand in devising a civilian workforce to handle the skill sets found lacking in the military force. Both the S/CRS and the PNSR are important elements to comprehend for this study; once we devise a strategy later in this book and identify stakeholders and funding streams, it will become clearer to the reader that both these initiatives are vital to our success in devising a framework and subsequent plan for conflict resolution.

DIPLOMACY, INFORMATIONAL, MILITARY, AND ECONOMIC—THE DIME FRAMEWORK

One of the most important lessons of the wars in Iraq and Afghanistan is that military success is not sufficient to win: economic development, institution-building, and the rule of law, promoting internal reconciliation, good governance, providing basic services to the people, training and equipping indigenous military and police forces, strategic communications, and more—these along with security, are essential ingredients for long-term success. Accomplishing all of these tasks is necessary to meet the diverse challenges in developing a war termination strategy that can be enduring.[69]

It is clear that Secretary of Defense Robert Gates believes that our national security structure must urgently devote time, energy, and thought to how we better organize ourselves to meet the international challenges of the present and the future. We need to ask, what institutions do we need for this post-Cold War world? "Harry Truman noted that if the Navy and Army had fought as hard against the Germans as they had fought each other, the war would have ended sooner."[70]

But it goes beyond the interservice rivalries that are at times responsible for the fog of war. The inability of this nation (or any nation for that matter) to get a "whole of government" approach to fighting its wars is arguably the single most problematic issue on the table today. When talking about the importance of his nation to partake in the significant issues in this hemisphere, an ambassador to the United States recently stated: either get a seat at the table or be part of the menu. The same holds true for those agencies that are not only interested but required to be part of the process in war termination and conflict resolution.

"America is at war." This is the first line in the 2006 NSS. But, is the nation at war? This is not the same nation that provided a nation of resources as seen in World War II; a war that lasted four years—less than one-half of the length thus far of the War on Terror. Therefore, while the 2006 version of the NSS states that it is a "wartime national security strategy," I will beg to differ.[71]

To help understand how the nation can support an effort in conflict by injecting more than just combat, or hard power, we must first understand the categories from which power is derived. These categories include political, military, economic, social, infrastructure, and information systems (PMESII) and diplomatic, informational, military and/or economic (DIME). These two concepts are key in warfare and operations that depend on achieving effects from the activities.

Effects-based planning calls for consideration of a broader set of options and a broader understanding of their potential impacts. DIME options are to be considered, along with their potential impacts on the PMESII environment. Given that the cause-effect relationships among these are not well understood, modeling these relationships and using them to forecast plausible outcomes is a challenging technical problem. Understanding each portion of DIME and how each supports the hard/soft/smart power theorems provides the strategist the necessary background for future chapters once we develop the framework.

The fog of postwar, as the title of this chapter states, can be overcome by understanding not only the war termination and conflict resolution endstates, but also the framework for what elements of power should be provided and a listing of the stakeholders responsible for the resources. The next chapter, War Termination Goes Webster, ensures the reader has a clear understanding of the terms necessary for universality across all stakeholders in the fight.

CHANGING THE FACE OF BATTLE? OR JUST A CHANGING PROBLEM?

Has the face of battle changed? Or is it just another era with all the principles of war intact? To the strategist tasked with defining a war termination and conflict resolution answer to a war, the answer to these questions matter.

"The U.S. military faces an era of enormous complexity. The complexity has been extended by globalization, the proliferation of advanced technology, violent transnational extremists, and resurgent powers."[72] Doctrine expert Frank G. Hoffman states that ". . . the evolving character of conflict that we currently face is best characterized by *convergence* [emphasis added]. This includes the convergence of physical and psychological, the kinetic and nonkinetic, and combatants and noncombatants . . . of greatest relevance are the converging modes of war. . . . Terrorism and conventional, criminal, and irregular warfare . . ."[73]

Cooperative security, stability operations, and irregular warfare missions require a better understanding of the complex operational environment—notably through rich contextual understanding of the factors affecting stability. Further assessment, policy, and planning need to consider factors associated with institutional performance (community organizations, government ministries, legal structures, etc.) based on how societies emerge, develop, and function, as well as attributes that provide resiliency and flexibility.

Today, stability experts are faced with a new environment in which the world is highly interconnected, change is very rapid, and threats are multifaceted; all of these pose very different challenges to the U.S. government. The current financial situation underscores both the rapidity and global extent of economic collapse, and it has exacerbated problems in other areas.

Solutions in one area can have first-, second-, and third-order effects in other areas; these effects can be both positive and negative. Global average food prices increased by more than 80 percent from 2005 to 2008, sparking food riots in Africa, Asia, the Middle East, the former Soviet Union, and Central and South America. Contributing factors are complicated; for example, shifts in food demands may be contributing to the food price increases, as increasing use of food crops for biofuels and increasing demand for protein-rich diets dramatically decreases efficient use of grain calories. Estimates of future water availability are alarming, and while earthquakes often impact manmade water management structures, reports also suggest that geophysical changes caused by large dams may have triggered earthquakes, including China's 7.9 magnitude earthquake on May 12, 2008, along a fault near the Zipingpu Dam and Reservoir that left 80,000 people missing or dead. In addition to food and water issues, severe weather events and climate change, shifts in demographics, increasing energy demands, pandemics, and threatened usage of nuclear weapons are perils that individually, or worse, in combination, can significantly increase the fragility of world stability.

While in today's increasingly interconnected world, global crises and unstable regions pose an acute risk to world security and could provide unforeseen circumstances ripe for manipulation and exploitation, these same threats can serve as rallying instruments, catalyzing disparate groups to work in concert to develop coordinated responses and preparedness mechanisms. This coordination can also result in development of negotiation venues for other issues. First-, second-, and third-order effects can have positive impacts. In addition to the factors identified in the frameworks above, other dimensions of consideration include (1) institutional performance (community organizations, government ministries, legal structures, etc.) as a function of how societies emerge, develop, and operate, and (2) attributes that provide resilience and flexibility. So the lingering question remains, where will the next occurrence of regional instability be that requires U.S. intervention? How should we shape the structure of our future force to respond to such instability? Examining the following aspects of this challenging problem include:

- How individual factors impact regional stability
- How these individual factors can combine to create tipping points that drive significant regional instabilities
- How understanding approaches to forecast and anticipate where future instabilities may occur can give our warfighters an advantage in the future.[74]

CHAPTER 2

War Termination Strategy Goes Webster

Democracy is the worst form of government, except for all those other forms
that have been tried from time to time.
—Winston Churchill, House of Commons speech on Nov. 11, 1947

One may wonder why this chapter opens with a quote by Churchill disparaging democracy. Words matter. Democracy means different things to different people even though there is a definition that is widely accepted. The timing of this famous remark is significant. Churchill won the war, but in the election of July 1945, he was defeated. He blamed the freedom of the people, or democracy, for the loss.

U.S. Forces–Iraq Commander, General Raymond T. Odierno stated during the spring 2010 election in Iraq that legitimacy is important but if there is less than 50 percent turnout for a fledgling democracy's first election then it would probably have failed at "selling" democracy as a form of government.[1] Fortunately for U.S. Forces–Iraq, the March 2010 election reported over 60 percent. Although this election was "successful," it had overwhelming support from the U.S. military and the political arm of the U.S. Embassy. The outcome of the Iraqi elections in 2010 may have been the most legitimate ever for Iraq or for any other recently liberated dictatorship, but the problem is that the future of elections without the U.S. support is a concern. There were clearly flaws in election procedures, and that is with oversight from the UN, the United States, and international observers. What happens when the "occupiers" leave? Is democracy in this case vulnerable to a shelf life?

Democracy, in Churchill's case, provided an undesired outcome to a man who championed its cause for an entire lifetime. In the end, do we look at spreading democracy or just diversity? And in a fledgling democracy does a legitimate election (note I did not state "fair," for fairness in any new democracy is a stretch at best) define success?

Many involved in prescribing strategy for wars oftentimes unwittingly misuse the terminology and, therefore, add to the fog of postwar described in Chapter 1. In this chapter, I will attempt to cut through the ambiguities by provide a full array of definitions, not only to words used in this and various other studies, but also many of the thoughts derived over the years. In the end, this chapter's goal is to help the reader better understand the eventual methodology produced by this work.

The 19th century Prussian soldier, military leader, and military theorist Carl von Clausewitz wrote the now-famous statement: "No one starts a war, or rather, no one in his senses ought to do so, without first being clear in his mind what he intends to achieve by that war and how he intends to conduct it."[2] According to Clausewitz, wars are brought to an end for three possible reasons: (1) the inability to carry on the struggle (i.e., defeat); (2) the improbability of victory; and (3) unacceptable cost.[3] From the great military theorist sprung the very first formal definition of "war termination."

This is a brief chapter that allows the reader to understand, in layman's terms, that words mean something when it comes to defining measurable, endstates, goals, and objectives; a necessary ingredient when devising a war termination strategy. It is important to be clear in what it means to *end, terminate,* and *transition* wars through *conflict resolution* and what it means to develop an *exit strategy* that encompasses these key terms. Although this list is not all-inclusive, it is meant to be instructive to the reader so strategists and planners have a baseline for understanding terms. Note that these definitions are pulled from many military manuals, doctrinal theses, commercial and professional articles, blogs, books, and interviews. They are my interpretations of the definitions and by no means represent an official stance. Unless otherwise referenced, most of these definitions are derived from the official Department of Defense (DoD) Dictionary, *Joint Publication 1-02, DoD Dictionary of Military and Associated Terms,* as amended through October 31, 2009 (http://www.dtic.mil/doctrine/dod_dictionary/).

DEFINITIONS OF WAR TERMINATION STRATEGY TERMS

Civilian Stabilization Initiative[4]: The Civilian Stabilization Initiative is a $248.6 million request in the FY2009 budget submitted to Congress by the Administration, with the objective of strengthening civilian capacity to manage and implement reconstruction and stabilization (R&S) activities. The Initiative includes funds for the Civilian Response Corps (Active, Standby, and Reserve components) across eight civilian agencies.

Active: The Active component of the Civilian Response Corps is composed of United States Government (USG) staff trained and ready to deploy to the field in 48 to 72 hours. It provides rapid response capacity to assess the situation, design the USG response, and begin R&S implementation.

Standby: The Standby component is composed of current USG employees who have ongoing job responsibilities, but volunteer to be trained and available for deployments in case of need. Standby members are deployable within 30 days for up to 180 days.

Reserve: Members of the Reserve component would be volunteers from outside the federal government, but would become USG employees when mobilized. They would be fully trained and deployable in 45 to 90 days to provide sector-specific expertise.

In the 2008 Supplemental Appropriations Act, Congress provided initial funding for the Active and Standby components of the Corps. Fully funded in FY2009, the Civilian Stabilization Initiative created an Active component of 250 members, a Standby component of 2,000, and a Reserve component of 2,000 members.

Coercive diplomacy: This type of diplomacy implies negotiating and fighting at the same time and can include escalation as one of its measures.[5]

Conflict resolution: This is a long process that is primarily a civil action that may require military support. In the past few conflicts in the Middle East, the military has taken the lead on most activities associated with the categories defined in Chapter 1 that encompass conflict resolution.[6]

Counterinsurgency (COIN): There are a number of definitions that define the current trend in conflict; counterinsurgency operations. The DoD says it is "those military, paramilitary, political, economic, psychological, and civic actions taken by a government to defeat an insurgency."[7]

The State Department defines it as "comprehensive civilian and military efforts taken to simultaneously defeat and contain insurgency and address its root causes."

These two "official" definitions reflect the fact that countering an insurgency requires a strategy tailored to the particular nature of the insurgency. As discussed above, this entails a comprehensive assessment of the root causes as well as the tactics, techniques, and strategy of the insurgents. The goal of COIN warfare is that the indigenous military fights an internal enemy.[8]

Counterterrorism (CT): Again, there are a multitude of definitions. DoD defines it as "operations that include the offensive measures taken to prevent, deter, preempt, and respond to terrorism." It is important to denote that CT is different from COIN in that terrorists are not necessarily in a competition for control of the population against the local or regional governing authority. CT operations are thus offensive operations, focused less on a competition for governance and more on undermining and disabling the terrorist *network*. In some Commonwealth countries, there is a distinction made between *antiterrorism*, which is defensive in nature, and *counterterrorism*, with is more offensive, disruptive, or preventive, as described here.

The Department of Defense Directive (DoDD) 3000.07 defines CT as operations that include the offensive measures taken to prevent, deter, preempt, and respond to terrorism. Although there is little doctrine or curriculum related to the topic of CT, *FM 3.05.20* (formerly *FM 31-20*), *Special Forces Operations*, has included material on CT since 1977, and the Special Operations Schoolhouse continues to conduct several courses related to countering terrorism. The subordinate categories of CT are hostage rescue, recovery of sensitive material from terrorist organizations, and attacks against terrorist infrastructure. As indicated in Title 10, U.S. Code, CT, along with Unconventional Warfare (UW) and foreign internal defense (FID), has been a core activity for special operations forces since 1987.

As it did during the mid-1960s and mid-1980s, the Army has done an exceptional job of relearning, re-establishing and re-institutionalizing its capability in the irregular warfare (IW) realm, but at a significant cost. For this period of interest to succeed where previous ones have failed, the focus must remain on institutionalizing the subject as a valid peer to other military subjects. IW must become a mainstream topic of the profession of arms rather than merely a fringe specialty relegated to a select few. Conversely, it must not be regarded (by the few) as an elite discipline, with the attendant pejorative view toward other military disciplines. History has shown that insurgency and terrorism will remain a normal part of the spectrum of conflict, often requiring the application of military power to preserve or protect U.S. national interests. The new challenge for this millennium is not the threat posed by IW or even how the Army will meet the challenge but rather how the Army will prepare itself for long-term success.[9]

Culminating Point: The culminating point is a point of equilibrium between the attack and the defense, a point at which the offensive can proceed no further without becoming relatively weaker than the defending force. The concept is certainly central to decisions on escalation, and the commander envisioning a negotiated settlement might well use the concept as a basis for planning war termination. Negotiations are unlikely to commence before the course of the war has become fairly clear, and an opponent who views the war situation favorably is more likely to be interested in negotiation than one who believes he is temporarily at a disadvantage.[10] The culminating point might be the time both sides are ready to negotiate.[11]

Deliberate Planning: The planning process is a cyclic process carried out in peacetime to develop and refine plans to be used in wartime. It is a detailed, intricate, five-phase process that can take 18 to 24 months. Yet nowhere in the chapters on deliberate planning is the critical nature of war termination criteria discussed.[12]

Economic Warfare: The planned use of economic measures designed to influence the policies or actions of another state, e.g., to impair the war-making potential of a hostile power or to generate economic stability within a friendly power. The employment of sanctions against another nation, such as in Iraq for most of the 1990s, can cause great degradation in a nation's infrastructure that deciding to occupy after an invasion could have lasting implications in conflict resolution.

Endstate: This is what the leadership desires the battlefield or operating environment and the surrounding political landscape to look like when the war is over. It the aim of the policies and strategies employed in a fight; the goals of national policy.[13] In short, the endstate is the set of required conditions that defines achievement of the commander's objectives.

Escalation: This is the deliberate move to higher or different forms of force, is central to limited war theory, and is a principal instrument in bargaining that for many theorists constitutes the war termination process. Escalation is an inherently interactive process, with all the uncertainties that it implies. Determining the forms and methods of escalation, calculating effects, coordinating it with diplomatic actions, and keeping it under control are essential tasks of war termination strategy.[14]

Foreign Internal Defense (FID): DoD defines it as "participation by civilian and military agencies of a government in any of the action programs taken by another government or other designated organization to free and protect its society from subversion, lawlessness, and insurgency." See also internal defense and development (IDAD), which emphasizes the preventative steps a state takes to protect itself from such threats.

As troop levels in Vietnam began to draw down in 1970, interest in COIN doctrine began to wane. The result was that the doctrine of the 1970s retained the COIN lessons learned in Vietnam and reflected the topic as military assistance to allied partner nations. In 1977, a chapter on FID replaced the chapter on COIN in *FM 31-20, Special Forces Operations*. Since the development of the FID concept, The Special Warfare Center and Schools (SWCS) has remained the proponent for its doctrine. The Military Assistance Training Advisor (MATA) Course, which began at the Special Warfare Center in 1962, trained joint military personnel in the skills required to serve as advisers, predominantly in South Vietnam. The training included language instruction that was similar to that of the UW course. Although the MATA course closed in 1970, many of its lessons were retained and incorporated into the special forces, psychological operations, and civil affairs courses. The topic of FID proved to be so valuable that in 1994, SWCS produced the first FID field manual. That eventually led to the development of *JP 3.07.1, Joint Tactics, Techniques, and Procedures for Foreign Internal Defense*, which was written for the joint military community by the U.S. Special Operations Command.[15]

Grand Strategy: The word *strategy* pervades American conversation and our news media. We tend to use strategy as a general term for a plan, a concept, a course of action, or a "vision" of the direction in which to proceed at the personal, organizational, and governmental—local, state, or federal—levels. Such casual use of the term to describe nothing more than "what we would like to do next" is inappropriate and belies the complexity of true strategy and strategic thinking. It reduces strategy to just a good idea without the necessary underlying thought or development. It also leads to confusion between strategy and planning, confining strategic possibilities to near-time planning assumptions and details, while limiting the flexibility of strategic thought and setting inappropriately specific expectations of outcomes.[16] Instead, grand strategy, simply put, is the purposeful employment of all instruments of power available to a security community.

Guerrilla Warfare (GW): DoD defines GW as "Military and paramilitary operations conducted in enemy-held or hostile territory by irregular, predominantly indigenous forces." In fact, *guerrilla* is derived from the Spanish word for war, *guerra*, with the suffix *illa* meaning "little."

Hybrid Warfare: Multimodal or multivariant conflict that isn't simply black and white but blurring and blending of war.[17]

Internal Defense and Development (IDAD): DoD says that IDAD is "The full range of measures taken by a nation to promote its growth and to protect itself from subversion, lawlessness, and insurgency. It focuses on building viable institutions (political, economic, social, and military) that respond to the needs of society."

IDAD is a term that has been used since the Cold War and in many respects can be seen as the more preventative side of COIN or FID. COIN is conducted in response to, rather than in anticipation of, an insurgent threat (though early intervention is preferred). FID emphasizes the role of an intervening power in supporting the security elements of a state.

Insurgency: DoD says an insurgency is "an organized movement aimed at the overthrow of a constituted government through use of subversion and armed conflict." The State Department offers a bit of a twist in stating that it is "the organized use of subversion and violence to seize, nullify, or challenge political control of a region."

Interagency Management System for Reconstruction and Stabilization (IMS):[18] IMS is designed to provide policy makers in Washington,

Chiefs of Mission (COMs), and military commanders with flexible tools to achieve:

- Integrated planning processes for unified USG strategic and implementation plans, including funding requests;
- Joint interagency field deployments; and
- A joint civilian operations capability including shared communications and information management.

The IMS is composed of the Country Reconstruction & Stabilization Group (CRSG), the Integration Planning Cell (IPC), and the Advance Civilian Team (ACT).

The Country Reconstruction & Stabilization Group leads CRSG policy formulation. The CRSG consists of a Washington-based interagency decision-making body, supported by a full-time interagency Secretariat that performs planning and operations functions and mobilizes resources. The CRSG is co-chaired by the Regional Assistant Secretary of State for the country in question, the S/CRS Coordinator, and the appropriate National Security Council Senior Director.

- **The Integration Planning Cell (IPC):** IPC consists of interagency planners and regional and sectoral experts who deploy to the relevant Geographic Combatant Command or multinational headquarters to assist in harmonizing ongoing planning and operations between civilian and military agencies and/or the USG and multinational headquarters (HQ).
- **Advance Civilian Team (ACT):** ACT supports the Chief of Mission (Ambassador) in the field to develop, execute, and monitor plans. ACT provides interagency field management, deployment, and logistics capabilities, developing and implementing activities through regional field teams.

Irregular Warfare (IW): DoD defines it as "A violent struggle among state and nonstate actors for legitimacy and influence over the relevant population(s). Irregular warfare favors indirect and asymmetric approaches, though it may employ the full range of military and other capacities, in order to erode an adversary's power, influence, and will."

This term is increasingly popular in the U.S. military, especially the Marine Corps and the special operations community. Others, including many in the U.S. Army, object to the term because of its conceptual confusion and because nonmilitary partners object to having their mission sets re-crafted under the "war" terminology. This is especially the case among diplomats who understand that with respect to diplomacy and international law, *war* has very specific meanings. Other objections in the military community focus on the intellectual difficulty in categorizing too many things under one "umbrella" term (such as "MOOTW") and in defining something by what it is not (i.e., not "regular").[19]

Nation Building: Nation building refers to the process of constructing or structuring a national identity using the power of the state. This process aims at the unification of the people or peoples within the state so that it remains politically stable and viable in the long run. Nation-building can involve the use of propaganda or major infrastructure development to foster social harmony and economic growth. The important point to capture here is that the form of nation-building and resources required depends on the starting point. If what had previously existed has been destroyed or if there was never really a nation (much like Somalia) to begin with, then the effort and activities involved in nation-building will be great.

The Clinton Administration's experience in the Balkans in nation building is an example of how I will describe the definition of nation-building. The Clinton Administration became more capable over the eight-year tenure by learning. However, it was not the same for Bush—and this is a major *however*—because Bush had to endure 9/11, and regardless of what some would call the heroic amateurish Coalition Provisional Authority (CPA), Bush had a different hand to deal with.

NSPD 44: The president issued National Security Presidential Directive 44: Management of Interagency Efforts Concerning Stabilization and Reconstruction, on December 7, 2005, in response to the recognized need for whole-of-government planning and response to crises abroad. The goal of NSPD 44 is to promote the security of the United States through improved coordination, planning, and implementation of stabilization and reconstruction assistance. NSPD 44 empowers the Secretary of State to lead and coordinate the U.S. government response across all involved agencies, and to work with the Secretary of Defense to harmonize civilian and military activities. The Office of the Coordinator for Reconstruction and Stabilization (S/CRS) in the Department of State (DOS) facilitates this.

Power: The dictionary tells us *power* is the capacity to do things. This is very important for nowhere does it say "military might" but rather the ability to get the outcomes one wants. It also says it is to have the capabilities to affect the behavior of others to make those things happen or to influence others.

Because to get desired outcomes requires necessary resources, *power* can also mean the possession of capabilities or resources that can influence outcomes. Therefore, a nation is powerful if has a great population and resources in the terms of economic, military and social strengths. But this definition, as stated Nye in his seminal work *Soft Power*, "power resources are not as fungible as money." Converting resources into realized power, as Nye says, in the sense of obtaining desired outcomes requires well-designed strategies and skillful leadership.[20] But the distribution of power and resources varies even inside a nation's own administration. Clearly the disparity between the resources in the DoD and the Department of State (DoS) are striking in their difference. The DoS budget in 2008 was 2 percent of the entire DoD budget. There is also a difference in capacity. The interagency partners do not have the capacity of the US military. DoS is far smaller in both size and budget than DoD. The DoS, together with the United States Agency for International Development (USAID), has about a $30 billion budget and 57,000 employees of which half are foreign nationals. This is in stark contrast to DoD with its $580 billion budget and 3 million work force. Clearly the disparity of soft and hard power is obvious when we define this type of difference in resources[21] (see Figure 2.1 for a budget comparison for the collusion of governmental agencies; better known as the Interagency).

Stability operations (SO): SO is an overarching term encompassing various military missions, tasks, and activities conducted outside the United States in coordination with other instruments of national power to maintain or re-establish a safe and secure environment and to provide essential governmental services, emergency infrastructure reconstruction, and humanitarian relief. These operations are military and civilian activities conducted across the spectrum from peace to conflict to establish or maintain order in states and regions. These operations are executed outside the

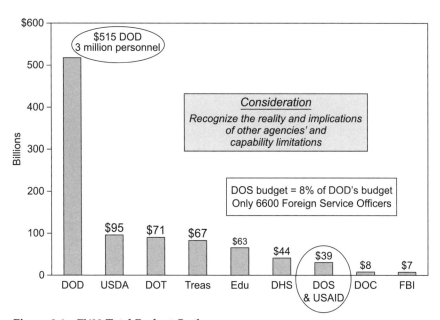

Figure 2.1 FY09 Total Budget Outlay

United States and its territories to establish, preserve, and exploit security and control over areas, populations, and resources. They may occur before, during, and after offensive and defensive operations; however, they also occur separately, usually at the lower end of the spectrum of conflict. Army Forces engaged in stability operations establish, safeguard, or restore basic civil services. They act directly in support of government agencies. Throughout our history, the Army has regarded stability operations as "someone else's job," an unwanted burden—a series of sideshows that soldiers performed either separately from war or in the wake of war. Therefore, such operations were belittled for diverting essential resources away from the military service's principal mission of warfighting. Conceptually, they lead to an environment in which the other instruments of national power can predominate.[22]

Stability operations, like FID, saw doctrine developments as interest in COIN lessened following Vietnam. Much COIN doctrine was incorporated into the stability operations chapter of *FM 31-20*, as well as into the psychological operations, special forces, and civil affairs courses. In 1967, the handbook from the MATA course was used to develop *Field Manual 31-73, Handbook for Advisors in Stability Operations*.[23]

In the past, the following kinds of activities have fallen under the rubric of stability operations:[24]

- Peace operations
- Foreign internal defense
- Security assistance
- Humanitarian and civic assistance
- Support to insurgencies
- Support to counterdrug operations
- Combating terrorism
- Noncombatant evacuation operations
- Arms control
- Show of force

Strategy: Doctrinally, strategy is a prudent idea or set of ideas for employing the instruments of national power in a synchronized and integrated fashion to achieve theater, national, and/or multinational objectives. In simplistic terms, strategy at all levels is the calculation of objectives, concepts, and resources within acceptable bounds of risk to create more favorable outcomes than might otherwise exist by chance or at the hands of others. Strategy is defined in *Joint Publication 1-02* as "the art and science of developing and employing instruments of national power in a synchronized and integrated fashion to achieve theater, national, and/

or multinational objectives."[25] What is key here is the identification of employing the instruments of national power. When it comes to conflict and all-out war, most think that strategists focus solely on the hard power, or military might. The operating environment today and all its complexities require an expansion of this to other forms of power.

Victory: The great strategist B.H. Liddell Hart wrote that, "If you concentrate exclusively on victory, with no thought for the aftereffect, you may be too exhausted to profit by the peace, while it is almost certain that the peace be a bad one, containing germs of another war."[26] This is profound because strategists and tacticians often lose sight of the fact that closing with and destroying the enemy only leads to an aftermath; that aftermath is conflict resolution and an ultimate endstate.

"Victory may be the proper object for a military campaign but rarely for a war."[27] Victory doesn't have to be total annihilation a la World War II Soviet Blitzkrieg or resemble the typical American inherited tradition of "total" victory in conflicts defined between good and evil.[28] Although sometimes fighting becomes an end it itself, its termination is an essential link between national strategy and posthostility aims and that military victory is measured by how it supports overall political objectives.[29]

War Termination: Military doctrine states ". . . conflict termination should be considered from the outset of planning."[30] *FM 3-0*, for example, uses the terms *conflict termination* and *war termination* within its pages, but defines *conflict termination* as "the point at which the principal means of conflict shifts from the use or threat of force to other means of persuasion."[31]

Griffith observed that "one or possibly both nations must adjust aims if the war is to be terminated. If the original aims were mutually attainable, the war would have been unnecessary."[32] It is always easier to get into a conflict than to get out of one. Current joint doctrine recognizes that although coercive military operations may end, the conflict may continue under other means such as terrorism, insurgency, cyber war, economic disruptions, political actions, or acts of civil disobedience.

War termination is the process leading to the resolution of a conflict and the basis for mutual acceptance of interests and objectives to ensure lasting settlement conditions. War termination not only includes the use of force but may involve all the instruments of power such as political, economic, and information:[33]

Specific goals may not necessarily translate into preplanned exit strategies or a clearly identifiable conflict termination.[34] All of the terms are used universally as if there is common understanding of each. However, this is not the case. Exit strategies are highly desirable because the lack of one could result in reduced confidence in leadership, drop in troop morale, possibility of increased casualties, and possible negation of any successes achieved by the actual intervention which, in turn, may diminish public support.[35] War termination can cover a lot, from planning before a war starts to the negotiations following the cessation of violence. Narrowly construed, war termination might be the process of deciding when and how to stop the fighting when it becomes evident that war fighting objectives have been met or perhaps, are no longer achievable. The most telling thing about the term *war termination* is that it isn't even recognized as a term worthy of definition in *Joint Staff, J-7, Joint Publication 1-02, Department of Defense Dictionary and Associated Terms,* which confirms the need for this book.

Whole of Nation (WoN), Whole of Government (WoG), and **Whole of World (WoW)** or **Whole of Nation/Government/World (WoN/G/W):** Reconstruction and stabilization planning is undertaken in support of achieving *conflict transformation* in the specified country or region. The goal of conflict transformation is to reach the point where the country or region is on a sustainable positive trajectory and where it is able to address, on its own, the dynamics causing instability and conflict. This requires simultaneously building local institutional capacity while reducing the sources of instability and conflict—all during the two- to three-year window of opportunity when resources and political will are most available. One fundamental principle of conflict transformation is that, over the longer term, the host nation must develop its own capacity to ensure stability and conditions for economic growth—those conditions cannot be imposed from outside.[36]

HIERARCHY OF NATIONAL DOCUMENTS AND UNDERSTANDING

It is important to go into some detail on the U.S. national security strategy (NSS) and national military strategy (NMS) if one is to understand a nation's (at least the United States's) focus on a positive endstate or goal when getting into a fight. The NMS is a document approved by the Chairman of the Joint Chiefs of Staff for distributing and applying military power to attain NSS and national defense strategy objectives. The NSS is a document approved by the president of the United States for developing, applying, and coordinating the instruments of national power to achieve objectives that contribute to national security. Both the NSS and NMS and, to some degree, former Secretary of Defense Donald Rumsfeld's National Defense Strategy (NDS) have driven policy and strategy in the past decade. The evolution of the NDS was, in essence, because of the State Department's and to some degree the NSA's inability to devise a WoG policy to drive strategy. In short, the NDS, although not part of the methodical way of developing a processed nested national security and military strategy, ensure endstates and resources are developed and executed in a synchronized way.

The 2005 NDS was noteworthy for its expanded understanding of modern threats. Instead of the historical emphasis on conventional state-based threats, the strategy defined a broadening range of challenges including traditional, irregular, terrorist, and disruptive threats. The strategy outlined the relative probability of these threats and acknowledged America's increased vulnerability to less conventional methods of conflict. The strategy even noted that the DoD was "over invested" in the traditional mode of warfare and needed to shift resources and attention to other challengers.

Although civil and intrastate conflicts have always had a higher frequency, their strategic impact and operational effects had little impact on Western military forces, and especially U.S. Forces, which focused on the significantly more challenging nature of state-based threats and high-intensity conventional warfighting. This focus is partly responsible for America's overwhelming military superiority today measured in terms of conventional capability and its ability to project power globally. This investment priority and American force capabilities will have to change, however, as new environmental conditions influence both the frequency and character of conflict.[37]

On March 16, 2006, the latest NSS was issued. It is a return to the more multilateral approach of previous administrations. It restates America's commitment to supporting democracies and defeating terrorism, puts forth a plan to restructure institutions related to national security, and discusses the challenges of globalization. However, it has been more than four years since the last NSS was published. As the fight in Iraq draws to a close and the stresses put on the command in Afghanistan to expand its footprint, a need to define the potential strategic challenges is needed.

A desire to fight the last war(s), which consisted of counterinsurgency fights, will be great. But to fight the last fight has always proven fruitless. Desert Storm perhaps was the last major success in that the United States didn't fight its last major war, Vietnam. In fact, in the plan for Desert Storm and its preparation, the strategists and leaders did all it could do NOT to fight another Vietnam; meaning not to get bogged down in a quagmire where objectives were ill-defined. However, Operation Iraqi Freedom (OIF) and Operation Enduring Freedom (OEF) were fought like the last big fight, the high-intensity conflict like Desert Storm where at the end of a few months of high-intensity conflict all involved were doing "high fives" on the battlefield, but unfortunately both OIF and OEF turned into a long drawn-out counterinsurgency.

The current U.S. leadership under the guise of President Barack Obama and the vision of his National Security Advisor, General (retired) James Jones, are expected to refine the National Security Agency (NSA) and its role in the interagency struggle to support the anticipated newly revised

NSS and nested NMS. These strategies must be specific in their desired endstate and not lack nesting along a hierarchy. Any ambiguous goal and objective connectivity between what the nation calls its strategy and how the military supports it with its robust resources will blur the lines when it comes to supporting preplanned efforts in defining the necessary potential war termination and conflict resolution requirements.

By thoroughly examining the structures and processes of the current legacy national security system—including its human and physical capital and management dimensions, as well as its executive–legislative branch dynamics—I have isolated the system's essential problems. Unless these essential, underlying problems are rectified, system failures will occur with increasing frequency. Five interwoven problems, which the report details at length, are key:

1. The system is grossly imbalanced. It supports strong departmental capabilities at the expense of integrating mechanisms.
2. Resources allocated to departments and agencies are shaped by their narrowly defined core mandates rather than broader national missions.
3. The need for presidential integration to compensate for the systemic inability to adequately integrate or resource missions overly centralizes issue management and overburdens the White House.
4. A burdened White House cannot manage the national security system as a whole to be agile and collaborative at any time, but it is particularly vulnerable to breakdown during the protracted transition periods between administrations.
5. Congress provides resources and conducts oversight in ways that reinforce the first four problems and make improving performance extremely difficult.

Taken together, the basic deficiency of the current national security system is that parochial departmental and agency interests, reinforced by Congress, paralyze interagency cooperation even as the variety, speed, and complexity of emerging security issues prevent the White House from effectively controlling the system. The White House bottleneck, in particular, prevents the system from reliably marshaling the needed but disparate skills and expertise from wherever they may be found in government, and from providing the resources to match the skills. That bottleneck, in short, makes it all but impossible to bring human and material assets together into a coherent operational ensemble. Moreover, because an excessively hierarchical NSS is too bureaucratic and myopically stovepiped as a whole, it cannot achieve the necessary unity of effort and command to exploit opportunities.

The resulting second- and third-tier operational deficiencies that emanate from these five basic problems are vast. As detailed in the report, among the most worrisome is an inability to formulate and implement a coherent strategy. Without that ability, we cannot perform even remotely realistic planning. The inevitable result is a system locked into a reactive posture and doomed to policy stagnation. Without a sound strategy and planning process, we wastefully duplicate efforts even as we allow dangerous gaps in coverage to form. These systemic shortcomings invariably generate frustration among senior leaders, often giving rise to end runs and other informal attempts to produce desired results. Sometimes these end runs work as short-term fixes; other times, however, they produce debacles like the Iran-Contra fiasco.

A key part of the system's planning deficit arises from the fact that it is designed to provide resources to build capabilities, not to execute missions. Since we do not budget by mission, no clear link exists between strategy and resources for interagency activities. As things stand, departments and agencies have little incentive to include funding for interagency purposes; they are virtually never rewarded for doing so. As a consequence, mission-essential capabilities that fall outside the core mandates of our departments and agencies are virtually never planned or trained for—a veritable formula for being taken unawares and unprepared.

This explains why departments and agencies, when faced with challenges that fall outside traditional departmental competencies, almost invariably produce ad hoc arrangements that prove suboptimal by almost every measure. Personnel are often deployed to missions for which they have little if any relevant training or experience. It also explains why in novel environments, such as nation-building missions in Iraq and Afghanistan, multiple U.S. departments and agencies have trouble cooperating effectively with each other; nothing has prepared them for it.

An overburdened White House also produces an array of less obvious collateral damage. As a rule, U.S. presidents have resorted to two means of reducing their burdens when the interagency process fails to produce adequate policy integration: designate a lead agency or a lead individual—a czar. Neither means has worked well. Neither a lead organization nor a lead individual has the de jure or de facto authority to command independent departments and agencies. The *lead agency* approach thus usually means in practice a *sole agency* approach. Similarly, czars must rely on their proximity to the president and their powers of persuasion, which, if institutional stakes are high, can be downplayed if not entirely dismissed. The illusion that lead agency or lead individual fixes will work in turn tends to demobilize continuing efforts at creative thinking among senior officials, thus enlarging the prospect of ultimate mission failure.

White House centralization of interagency missions also risks creating an untenable span of control over policy implementation. By one count, more than 29 agencies or special groups report directly to the president. Centralization also tends to burn out National Security Council staff, which impedes timely, disciplined, and integrated decision formulation and option assessment over time. Further, time invariably becomes too precious to be spent rigorously assessing performance, which essentially vitiates any chance for institutional learning and dooms the system to making the same mistakes over and over again.

Lastly in this regard, the time pressure that an overburdened White House faces almost guarantees an inability to do deliberate, careful strategy formulation, thus completing the circle that ensures the system's inability to break out of its own dysfunctional pattern. When there are fires to put out every day, there is little opportunity to see and evaluate the bigger picture. Too short-term a focus also blinds leaders to the need to attend to system management and design issues. This significantly compounds the system's inability to learn and adapt.

The results are cumulatively calamitous. Without a realistic and creative national security strategy, no one can say what policy balances and tradeoffs are needed. No one can devise a rational investment strategy. No one can devise appropriate human resources and education programs to ensure an effective system for the future or recognize the critical importance of generating a supportive common culture among national security professionals.

Ossified and unable to adapt, our NSS today can reliably handle only those challenges that fall within the relatively narrow realm of its experience in a world in which the set-piece challenges of the past are shrinking in frequency and importance. We are living off the depleted intellectual and organizational capital of a bygone era, and we are doing so in a world in which the boundaries between global dynamics and what we still quaintly call domestic consequences are blurred almost beyond recognition. We thus risk a policy failure rate of such scope that our constitutional order cannot confidently be ensured.[38]

STRATEGY, GRAND STRATEGY, AND POLICY

Understanding how strategy, grand strategy, and policy fit in the hierarchical scheme of conflict planning and resolution is an imperative to this study. War termination strategies must work at the interface between grand strategy and military strategy—a difficult juncture where military art meets political policies.[39]

What will be instructive is the model grand strategy document known as *National Security Council Report 68* (NSC-68) that was, many would argue, the last grand strategic treatise this nation has seen. NSC 68 retains a talismanic significance, a model for what a "coherent grand strategy looks like.[40] NSC 68 was a 58-page classified report issued by the United States National Security Council on April 14, 1950, during the presidency of Harry S. Truman. Written during the formative stage of the Cold War, it has become one of the most significant historical documents of that era. NSC-68 would shape U.S. foreign policy in the Cold War for the next

20 years and has subsequently been labeled its "blueprint." It is not lost on today's policy makers and strategists that in an era with a single threat; the Soviet Union, NSC-68 was perhaps a somewhat simplistic and naturally uniting document to devise. Truman officially signed NSC-68 on September 30, 1950. It was declassified in 1975. Most would argue that with today's complex threat and non-nation-state challenges, an all-encompassing strategic plan is difficult to conceive.

Why is a definitive grand strategy so important? Putting out a fire is less efficient that preventing it from happening in the first place. To prevent a fire from igniting, proactive actions must be put in place. A nation must be proactive by having at least a plan on the shelf to devise a resolution for conflict as war comes to an end in a region. Oftentimes the lack of a proactive strategy leaves the forces involved in the close fight struggling to achieve success and *reacting* instead.

Former U.S. Joint Forces and the former NATO's Supreme Allied Transformation Commander General James N. Mattis observed that the United States lacks a grand strategy for the world it confronts today. However, he conceded, "It's hard to write a grand strategy when your house is on fire." Mattis admits that the military has overemphasized the employment of technology to solve strategy problems. "We have to diminish the idea of technology is going to change warfare.. . . War is primarily a human endeavor."[41]

Not having a comprehensive grand strategy since the fall of the Soviet Union has proven detrimental. Throughout the long, twilight struggle of the Cold War, the various parts of government did not communicate or coordinate very well with each other. There were military, intelligence, and diplomatic failures in Korea, Vietnam, Iran, Grenada, Bosnia, Kosovo, Haiti, and arguably today in Iran and Afghanistan. Getting the military and other services to work together is a recurring battle that had to be addressed time and time again, and was only really resolved by legislation in 1986.[42] When defining a grand strategy and what its conflict aftermath looks like, the key is to be able to clearly define both the political conditions and the situation that one envisions existing when both the conflict and dispute are over.

In *Strategy*, Hart analyzes the relationship between strategy, grand strategy, and policy. He argues that Clausewitz's definition of strategy as "the art of the employment of battles as a means to gain the object of war" has two defects. First, it blurs the distinction between strategy and policy, or between government, which is responsible for "the higher conduct of war," and the military leaders, which the government "employs as its agents in the executive control of operations." The second defect of the Clausewitzian formulation is its restrictive view of strategy: "It narrows the meaning of 'strategy' to the pure utilization of battle, thus conveying the idea that battle is the only means to the strategical [sic] end."[43]

With these two reservations in mind (as well as recognizing the definition of strategy as it is stated earlier in this chapter), Hart proposes that strategy be defined as "the art of distributing and applying military means to fulfill the ends of policy." It is important here to recognize that the "means," as defined by Hart, includes all other resources in *addition*, or perhaps, *in lieu of*, the employment of combat power. To be clear, the use of the term *military means* (in lieu of Clausewitz's "battles") broadens the meaning of strategy beyond actual fighting.[44]

At this stage, it is important to distinguish two meanings of *policy* as it relates to war. First, there is the policy that guides the conduct of war; second, there is the more fundamental policy that would govern its object. Once the shooting begins, war's innate dynamism takes center stage, and policy becomes hostage to military performance and changing domestic and international political circumstances.

Grand strategy is practically synonymous with the former meaning. Its role is to co-ordinate and direct *all* the resources of a nation or band of nations toward the attainment of the political object of war—the goal defined by fundamental policy. In relation to strategy, then, grand strategy is more encompassing, in terms of both the resources it uses and its horizon.

Four grand strategy categories that are common in the way the United States defines its stance in the international community are neo-isolationism, selective engagement, cooperative security, and primacy. [45] Perhaps the lack of a grand strategy in the post-Cold War era is mainly because the Friedman-type "flattening of the world" really is about *globalization*—which, in my opinion, is a submissive form of cooperative security. Regardless, to define any type of strategy that includes war and conflict termination attributes, we must first define our grand strategy. If we can define a grand strategy, perhaps as part of a newly devised NSS, then perhaps we can better define what wars or conflicts to engage and, for the purpose of this work, devise a plan for exiting and resolution prior to ever firing the first shot.[46]

On the latter, Hart writes that "whereas strategy is only concerned with the problem of winning military victory, grand strategy must take the longer view—for its problem is the winning of the peace." This difference in horizons implies that strategy sometimes has to be restrained for the sake of grand strategy, particularly when the pursuit of military decision— toward which the state may need to use all available force—results in self-exhaustion and a more bitter, resolute, and united opponent as seen in Southwest Asia.[47]

The components of grand strategy are as varied as its scope would suggest. Beyond the application of military means, Paul Kennedy's *Grand Strategies in War and Peace* focuses on three groups of factors: (1) scarce national resources, which constitute the means that must be balanced against the ends of policy; (2) peacetime and wartime diplomacy, which is concerned with "improving the nation's position—and prospects of victory—through gaining allies, winning support of neutrals, and reducing the number of one's enemies (or potential enemies)"; and (3) national morale and political culture, which govern the public's willingness to bear the costs and burdens of national security.[48]

One way of conceiving the structural relationship within grand strategy between these components and the military one has been suggested by Edward Luttwak: ". . . grand strategy may be seen as a confluence of the military interactions that flow up and down level by level [technical, tactical, operational, theater], forming strategy's 'vertical' dimension, with the varied external relations among states forming strategy's 'horizontal' dimension."[49]

Although this image does not maintain a clear-enough separation between action and outcome, it does call attention to the relationship between grand strategic design and the outcomes at the grand strategic level. The design stage is informed, of course, by expectations about outcomes. However, these expectations must be based on some conception of how grand strategic outcomes are produced and, more specifically, on how the interaction among various components of grand strategy operate in reality.[50]

To put this in perspective, Luttwak attributes the success of the North Vietnamese in the Vietnam War to a grand strategy that effectively compensated for weakness in the vertical (military) dimension with strength in the horizontal dimension: the military campaign, which was incapable of defeating American troops in battle. He writes that this grand strategy was designed to prolong the war so that diplomacy and propaganda could undermine public support for it in the United States and among its allies.

This perspective suggests that an explanation for the occasional puzzling defeat of great powers at the hands of minor ones is to be found at the grand-strategic level, where nonmilitary factors can sometimes compensate for inferiority in the hard currency of military resources. Thus, focusing only on the vertical dimension of strategy can be misleading in both predicting and accounting for conflict outcomes.

How misleading depends on the interrelationships among the various components of grand strategy and on a broader, systemic context in which it is applied. Military technology, communications, and norms are some of the factors that shape the context of grand strategy. When this context changes, grand-strategic success depends on decision makers' ability to acknowledge the new environment and adjust grand strategy accordingly. In recent years, the growing importance of the communications revolution has led in the United States and elsewhere to a re-evaluation of grand strategy and to a corresponding acknowledgment of the role that public diplomacy should play in it. Knowing all these factors or surmising them at the very least and how they are conveyed publicly both at home and internationally before committing forces is helpful. Knowing them and devising a termination and resolution plan based on them is even better in helping policy makers and strategist to envision achievable endstates.

PUBLIC DIPLOMACY AND STRATEGIC COMMUNICATIONS

An important part of recognizing an achievable endstate as it relates to a nation's grand strategy is the criticality of public diplomacy and strategic communications in developing an exhaustive plan for conflict and its resolution. History has shown myriad examples where policy is driven by the public and, in turn, the strategy evolves in such a way that an endstate is never fully defined. Starvation and poverty in Africa, for example, seems the catalyst oftentimes for public notice and international interest. The U.S. public invariably insists on saving the day, feeding the children, or suggesting intervention with fervor in bringing "hope" to a struggling populace. On the flip side, we often see a national uproar when these types of humanitarian tragedies are ignored by the national decision makers.

Instead of letting public opinion drive policy and the ensuing strategies, the strategists and politicians would be wise to use public diplomacy and strategic communications to drive public opinion. Often the public lets its heart strings drive its passion to demand to "just do something," and the result is politicians scrambling to act without thought for their constituents. What is lacking is a clear, concise endstate and related objectives. Clearly, given these conditions, there is little chance that a war termination phase in a quest to "just do something" can survive the planning phase. Jeffrey Record stated that, "Exit strategies, like Jominians everywhere, tend to discount the degree to which nonscientific, especially irrational (read emotional public), factors continue to influence the conduct of war, including postulation of war aims."[51]

Policies and strategic communication cannot be separate. Strategic communication requires a sophisticated method that maps perceptions and influences networks, identifies policy priorities, formulates objectives, focuses on "doable tasks," develops themes and messages, employs relevant channels, leverages new strategic and tactical dynamics, and monitors success. Strategic communications is clearly an important facet in the development of well-nested strategic endstates and objectives.[52]

A war termination framework that honors the requirements of public diplomacy and strategic communications will build on in-depth knowledge of other cultures and factors that motivate human behavior. It will adapt techniques of skillful political campaigning, even as it avoids slogans, quick fixes, and mind sets of winners and losers; for it is the campaigning ideas by leaders that sow the seeds of future conflicts. It will search out credible messengers and create message authority. It will seek to persuade within news cycles, weeks, and months. It will engage in a respectful dialogue of ideas that begins with listening and assumes decades of sustained effort. Just as importantly, through evaluation and feedback, an all-encompassing public diplomacy and strategic communications plan enables political leaders and policy makers to make informed decisions on changes in strategy, policies, messages, and choices among instruments of statecraft to use in potential conflict and the resolution of such.[53]

Nothing shapes U.S. policies and global perceptions of U.S. foreign and national security objectives more powerfully than the president's statements and actions, and those of senior officials. How the U.S. government communicates with the world—explaining policies, presenting facts about American life and values, promoting the national interest by helping foreign audiences understand America—is a matter of no small importance. During the Cold War, for example, engagement in the war of ideas through the United States Information Agency was a critical element of the West's victory over the Soviet Union. Today, new informational and ideological challenges to American leadership have arisen: radical Islamism, Chinese expansionism, Russian revanchism, Venezuelan disinformation, etc. Furthermore, the media environment has become far more competitive and diverse than was the case during the Cold War. If there was ever a time that called out for a new and sophisticated U.S. public diplomacy doctrine, this is it.[54]

Interests, not public opinion, should drive policies. But opinions must be taken into account whenever policy options are considered and implemented. At a minimum, we should not be surprised by public reactions to policy choices. Policies will not succeed unless they are communicated to global and domestic audiences in ways that are credible and allow them to make informed, independent judgments.

In a chapter of *Taken by Storm: The Media, Public Opinion, and U.S. Foreign Policy in the Gulf War*, Jarol Manheim defines public diplomacy as "efforts by the government of one nation to influence public or elite opinion in a second nation for the purpose of turning the foreign policy of the target nation of advantage."[55]

This formulation, which could also encompass strategic bombing or terrorism campaigns, omits the *persuasive* nature of the interaction, as well as its channel and form. Another influential definition by Hans Tuch in *Commanding with the World: U.S. Public Diplomacy Overseas* addresses these features, but loses Manheim's emphasis on intended foreign policy effects: "a government's process of communicating with foreign publics in an attempt to bring about understanding for its nation's ideas and ideals, its institutions and culture, as well as its national goals and current policies."[56]

There are many definitions and slight nuances, but it is important to understand how public diplomacy differs from traditional diplomacy. One of the earliest definitions of public diplomacy was articulated by the

Edward R. Murrow Center of Public Diplomacy at the Fletcher School of
Law and Diplomacy, Tufts University, in the 1960s:

> Public diplomacy . . . deals with the influence of public attitudes on the
> formation and execution of foreign policies. It encompasses dimensions
> of international relations beyond traditional diplomacy; the cultivation by
> governments of public opinion in other countries; the interaction of private
> groups and interests in one country with another; the reporting of foreign
> affairs and its impact on policy; communications between those whose job is
> communication, as diplomats and foreign correspondents; and the process
> of intercultural communications.[57]

This definition of public diplomacy explicitly mentions private groups
and the media, and implicitly expands the involved parties to those whose
job is communications. In today's global marketplace, business is undoubtedly the leading communicator when it comes to the volume, frequency,
and sophistication of communications with peoples around the world.[58]

The two previous definitions also reflect a common distinction between
public diplomacy's long-term, culture-based activities and its short-term,
current affairs focus. Regardless of definition, it is clear that public diplomacy plays an integral part in accomplishing a nation's, or even the international community's, abilities to see the grand strategic goal. This is an
integral part in devising conflict termination strategies.

The perspective of the Bush Administration on public diplomacy, as
developed by the Under Secretary for Public Diplomacy and Public Affairs,
Karen Hughes, is somewhat different from these two definitions. It lays a
foundation for public diplomacy based on three strategic objectives:

1. Offering people around the world a positive vision of hope and opportunity
   that is rooted in America's belief in freedom, justice, opportunity, and respect
   for all.

2. Isolating and marginalizing the violent extremists, undermining their efforts to
   portray the West as in conflict with Islam, and demonstrating respect for Muslim cultures and contributions to the world community.

3. Fostering a sense of common interests and values between Americans and people of different countries, cultures, and faiths throughout the world.

It is the view of Business for Diplomatic Action that a new definition
of public diplomacy should incorporate elements from all three of these
approaches. Specifically, a new definition should recognize that effective
public diplomacy depends on:

- A compelling and credible representation of America's values, vision, and voice
  to the world, demonstrating respect for all cultures;

- An acknowledgment that listening and dialogue lay at the heart of the public diplomacy process;
- The active involvement of nongovernmental actors such as the media, the business community, nonprofit organizations, and individual Americans. [59]

Indeed, it could be argued that this form of influence is particularly suitable to post-Cold War environment, where broad trends have merged to create favorable conditions for public diplomacy: the long-term trend for democratization that has nearly doubled in the 1990s. The significance of these trends includes the fact that public opinion has become a factor of increasing importance and weight in foreign policies of many states. Additionally, the emergence of the recognition of the importance of foreign public opinion on U.S. policies through the use of *soft power* vice the use of hard power; which is consistent with the role and status of persuasion in a democratic society. Finally, these trends have brought about the lessening dependence of citizens on their governments and the local press for information on foreign events, and have vastly increased the potential targets for the *direct* communication of diplomatic messages. All of these developments conform to and enhance the characteristic components of public diplomacy as a form of influence.

Without a doubt, these trends are not uniformly distributed across the international system and do not impinge equally on the foreign and domestic policies of all states. But no other set of norms rivals the democratic one as a medium of international discourse and justification; indeed, public diplomacy campaigns aimed at the developed world incorporate its values and norms. Likewise, the influence of the information revolution is not restricted to the developed world, as is evidenced even—and perhaps precisely—in the resistance it meets from some nondemocratic governments. Furthermore, one should not underestimate the exploitation of new technologies in third-world public diplomacy that is aimed at the developed world, especially in the United States.

Public relations were invented in the United States but we are terrible at communicating to the rest of the world what we are about as a society and a culture, about freedom and democracy, about our policies and our goals. "It is just plain embarrassing that al-Qaeda is better at communicating its message on the Internet than America."[60] The question becomes, how has one man in a cave managed to out-communicate the world's greatest communication society?

Every time an American bomb kills civilians in Afghanistan, the United States loses another battle in the information war to the Taliban. And despite more accurate weapons, more careful targeting, and speedier responses to the press, those that wield the hard power can't seem to figure out how to stop the setbacks in this decisive struggle for influence. However, a former top military official believes he may have the answer:

let the troops blog in combat, so they can ward off the accusations of atrocities as they fight. Perhaps breaking firewalls to let that happen or installing commercial networks throughout the battlefield will allow the natural progression of information flow; after all, the enemy's home page claims victory regardless of any outcome—ergo, regardless of fact, the world sees the enemy as winning—the blog or home page says so. The reason why this is important to this text is that it is part of the equation when measuring how to achieve an endstate. Even the most meticulously planned war termination and conflict resolution strategy can be affected by public opinion both abroad and at home.

The apparent abundance of civilian casualties in Afghanistan in May 2009, during a battle in western Afghanistan, is a good example of the impact on the cause-and-effect reaction that occurs during combat operations. Locals said 100 or more civilians might have died in the crossfire. In response, the U.S. military launched an investigation. Senior American officials hinted that the Taliban might have staged the whole thing—while the president and the secretary of state apologized for the loss of life. After a few days, investigators concluded that the civilian death toll was only about one-third of what was initially reported. But the damage was done: innocents were killed, the Americans looked blood-thirsty, and the Taliban notched another win in the campaign for hearts and minds. The Afghan president even demanded an end to American airstrikes.[61]

Former Air Force Secretary Michael Wynne stated that he thought the best solution might be to let the troops themselves document the story. "We need to make sure we capture the news cycle by providing our troops with something like a combat blogger." But that means changing the DoD's often-schizophrenic approach to bloggers in uniform. Within the armed services, there's a growing recognition that average soldiers are the most trusted voices the military has. But oftentimes military and civilian leaders are squeamish about letting their troops publish online.[62]

If the military allows bloggers to tell their own stories, then it would end these publicity nightmares. "This thing of letting the Taliban, letting Al Jazeera, letting the enemy public affairs unit get a hold of 24 to 48 hours of news cycle and then you announce that you're forming an investigative team—what is that?" Wynne says. "The sad part is, that when [the military] forms an investigative team, it looks like it's only for one reason: to cover it up."

In the end, the firsthand accounts are seen to have more authenticity. "If [that soldier] walked into a hut and blogged that there were 20 bad guys, they had 15 computers, 20 AK-47s—if he blogged that right away, even if it went to a command center—you'd be far better off than what we're doing now."[63]

During its war in Gaza earlier this year, the Israeli Defense Forces embedded combat cameramen in infantry units to defend often-controversial attacks on Hamas militants camping in schools and mosques. But the documentary tactic was obscured by brutal tactics, controversial weapons, and a larger communications strategy of indifference toward world opinion.

Similarly, U.S. soldier-bloggers won't be able to make much of a dent in the Afghan information campaign, if U.S. aircraft kill dozens of innocents on a regular basis. Battlefield actions speak louder than uploaded words.

But if the Taliban are manipulating the information environment to make U.S. attacks looks worse than they actually are, then the bloggers and photographers might be able to help. "If you take down a place in Afghanistan, you've got to be in there with your cameras before the bad guys unload the truck of bodies (if that's what they're doing). Or, if they've put a bunch of women and children in there so you can blow them up—you need to be in there first to know that," Wynne says. "We're saddened when . . . the President has to go on the record and say, 'we hate it when any casualties occur.'" Wynne adds. But the president "might not know" what actually occurred in such an incident—"because nobody blogged it."[64]

Matthew Currier Burden started a blogging Web page popular with military members and many Internet bloggers called *Blackfive* and wrote *The Blog of War,* a blogging anthology that is loaded with firsthand reports from the Internet diaries of soldiers in Iraq and Afghanistan. He wrote of the DoD's decision to gag military bloggers in 2007: "This is the final nail in the coffin for combat blogging. . . . No more military bloggers writing about their experiences in the combat zone. This is the best PR the military has—its most honest voice out of the war zone. And it's being silenced."[65]

CHAPTER 3

The Good

Thus it is that in war the victorious strategist only seeks battle after the victory has been won, whereas he who is destined to defeat first fights and afterwards looks for victory.

—Chinese Military Strategist Sun Tzu

This chapter presents examples of what went "right" as it pertains to planning for war termination, conflict resolution, measurements and criteria, endstate and policy aims, and in some instances, definitive exit strategies. In the following examples, it is instructive to review—perhaps by phase in the fight—what portions went well and use those lessons learned to help in the future when planning each stage of the fight in the ongoing wars such as Iraq and Afghanistan and for those thereafter.

Recalling from the Introduction, Michael Brown stated that "In the first twelve years of the post-Cold War (1990 to 2001), fifty-seven major armed conflicts took place in forty-five countries. In the first half of this period, the number of conflicts ranged from twenty-eight to thirty-three per year. Although the incidence of conflict dropped as the post-Cold War era stabilized, the number of conflicts has held steady since the late 1990s at around twenty-five conflicts per year."[1]

If one intends any conflict to advance long-term interests, one must consider the essential question of how the enemy might be forced to surrender or, failing that, what type of bargain might work to terminate the war. Such questions combine both the political and military realms. Not only the military contest but also domestic and foreign policy developments contribute to the war's outcome. Although the question of terminating a war should arise as soon as the war has begun or in the course of advanced planning, it tends to receive little or no attention in war plans.[2]

WORLD WAR II

The occupation of Germany after World War II is perhaps the paradigm of a successful postcombat operation in modern American history. After four years of bitter fighting and some initial friction between American combat units and the defeated German population, the U.S. Army shifted from its combat missions and literally reorganized and retrained its forces for their new peacetime role. The U.S. Constabulary in Europe effectively bridged the gap between the victorious Allies and the defeated populace by providing aggressive law enforcement, border control, and assistance to the Germans in rebuilding their country's infrastructure. The presence of U.S. soldiers also served as a symbol of the United States' resolve to reconstruct a devastated Germany and help shape it into the trusted friend and ally that it is today.[3]

Initially, tactical units deployed across the region to maintain order and security while the Germans began the arduous process of rebuilding their country. The challenges were daunting in the face of the massive destruction caused by the war and the flood of refugees into the Zone of Occupation. Redeployment of U.S. Forces earmarked for the Pacific theater and the demobilization of units returning to the United States compounded the challenges faced by U.S. Forces. The U.S. Constabulary was formed in the summer of 1946 to serve as the dedicated occupation and law enforcement element. With specialized training and distinctive uniforms, these soldiers assisted the rebuilding of the German law enforcement agencies, countered the black market, and controlled the international and interzonal borders. The post-World War II occupation of Germany, spanning almost 11 years, was a massive and diverse undertaking involving Britain, France, and the Soviet Union and, to varying degrees, a multitude of U.S. government departments and agencies. Moreover, the occupation was a major event in German history and in the history of the postwar world.[4]

In the Pacific theater, following Japan's defeat, the U.S. military occupied that country in conjunction with the terms identified in the Potsdam Agreement. By those terms, the two primary goals of the occupation of Japan were the complete demilitarization of the Japanese military complex and the democratization of Japanese society. U.S. Army General Douglas MacArthur was appointed as the Supreme Commander for Allied Powers (SCAP) and was responsible for administering the Japanese occupation.

The plan for the occupation of Japan was known as Operation Blacklist. The plan was grounded in a planning document dating back as early as 1941. The two phases of the occupation included a brief initial phase that lasted approximately 60 days and a longer phase devoted to political and economic reform that lasted several years. A key part, which the U.S. Government could have used in Operation Iraqi Freedom-1 (OIF 1), was the early decision to work through the Japanese government and administrative organizations, rather than to dismantle and replace them.

Pursuant to this decision, the following divisions were established to oversee existing Japanese government bureaus: government, economy and science, natural resources, public health and welfare, civil intelligence, legal, civil information and education, civil property custodian, and diplomacy. These divisions were staffed by U.S. civil servants and former military officers—approximately 3,500 of them at the peak in 1948; so, clearly, it was military-heavy as were all postconflict manning up until then and since.

Guided by MacArthur's policies, the new Japanese government instituted many far-reaching changes in Japanese politics, society, and culture. Economic reforms imposed on the Japanese government included democratization of economic opportunity, land reform, and expansion of the scope and quality of workers' rights. So successful was the Allied Forces' occupation policy in Japan that the Japanese Diet approved a new constitution for the nation in 1947. The U.S. occupation of Japan formally ended on April 28, 1952, after Japan and representatives of the 46 Allied nations signed a peace treaty in San Francisco on September 8, 1951.[5] The difference is that the U.S. Government committed more than 3,000 civilians, albeit the majority of whom had military experience, to the cause.

PANAMA

The one-term Bush administration launched two major military interventions in the late 1980s and early 1990s. The first in 1989 was against the regime of Manual Antonio Noriega in Panama, and the second, not even a year later, was to reverse Iraq's conquest of Kuwait. We will discuss later the fact that just months before he left office, Bush ordered troops into Somalia. While in office, George Bush saw to completion the termination of both Panama and the Gulf War fights, but the Somalia end strategy was left to his successor, Bill Clinton. All three conflicts ended differently; two, what I call "good" and one "bad." The leadership during each obviously played a big role. The planning for conflict termination following the war termination of each had the most significant impact in the end.

Operation Just Cause, the invasion of Panama in December 1989 that destroyed the Panama Defense Force (PDF), was the largest use of U.S. Forces since the Vietnam War—25,000 troops were committed. A swift and clear military success ensued after a forced entry by special operations and airborne forces.

This conflict was motivated by deterioration in U.S. relations with Noriega, beginning in the mid-1980s, that culminated in Noriega-inspired attacks on U.S. military personnel and their dependents in Panama. Both the Reagan administration and then the Bush administration tried to remove Noriega (the desired endstate) by means short of force including economic sanctions; highly publicized U.S. military exercises in the Canal Zone; a February 4, 1988, federal court indictment of the Panamanian dictator on a host of drug trafficking charges; and support, albeit weak, of an attempted coup against Noriega on October 3, 1989, by one of his henchmen.

These measures, however, served only to increase Noriega's defiance, driving him into such increasingly reckless behavior as securing from Panama's puppet legislative body a declaration of war on the United States and encouraging increased attacks by goon-squad "dignity battalions" on exposed American citizens. The event that caused the decision to invade was the murder of Marine Corps lieutenant Robert Paz and the detention of a U.S. Navy officer and sexual abuse of his wife. President Bush, determined to protect the lives of U.S. citizens, directed the U.S. Forces in Panama to bring General Noriega to justice in the United States.[6]

The United States also sought to "defend democracy" in Panama, to combat drug trafficking, and to protect the integrity of the Panama Canal treaty.[7] In the end, Noriega was removed, the government replaced by a more progressive leadership; the ancillary objectives of defending the canal's integrity, establishing a footprint to fight drug trafficking, and pursuing democracy in the Central American country all came to a resolute end. This is an example of a well-thought-out conflict with definitive endstate aims and war termination criteria (catch Noriega and nullify the Panama's Defense Forces' (PDF) capability to continue to fight) and a conflict resolution plan—replace the standing government with a fledgling democracy that is at peace with its neighbors and part of the international community.

However, by some reports Operation Just Cause caused heavy destruction on Panama and its citizens. Physicians for Human Rights (PHR) reported that "approximately 85 percent of the Panamanian lives lost during the invasion and its violent aftermath were civilian."[8] Approximately 300 civilians died while 50 military personnel were killed during the invasion. The U.S. government stated that approximately 1,000 Panamanians were injured, while PHR placed the number closer to 3,000. Whether the United States should have invaded Panama is not a question for this text, however, in general, it appears that while other alternatives were conceived and implemented, they did not properly affect the right people—those beneath Noriega who were tired of his iron-fisted and narcissistic rule. Noriega appears to have been one of the many despots of the 20th century who seemed to respond only to force and little else. In the end, the U.S. Government's consensus to use direct military force proved successful in dealing with Noriega in a finalizing way.

Much of the stability and reconstruction operations during and after Operation Just Cause were part of Operation Promote Liberty. This plan concentrated on public safety, health issues, and population control measures, as well as helping the newly appointed government of Panama function efficiently. An important part of the effort was to train the Panamanian police and paramilitary forces that replaced the PDF; an effort that was to be replicated in orders of magnitude in years to come in Iraq and Afghanistan. The primary organization responsible for executing this plan was the military support group, backed by U.S. Forces already stationed in Panama. In January 1991, the group was deactivated; its mission accomplished.[9]

FIRST GULF WAR

Operation Desert Storm, better known at the time as the "the Gulf War," was an overwhelming one-sided and clear coalition victory. "The Gulf War is cited by exit-strategy proponents as proof that control of a war's purpose and duration is possible."[10] Prior to launching the occupation of Saudi Arabia and what is known as Operation Desert Shield, the U.S. objectives established were clear, feasible, and then sustained without alteration throughout the Gulf Crisis of 1990 to 1991. This conflict, while limited and somewhat simplistic because of its massive supportive Coalition, is a model for war termination strategy and should be emulated in future planning. And while it is noticeably a model for war termination strategists, it cannot be lost on those determined to plan appropriately for the aftermath of conflict that Desert Storm was more than inadequate for establishing conflict resolution parameters. In short, the sanctions brought on by the United States after the Gulf War left the nation of Iraq a broken, poor, and incapable country. Those desperately trying to establish a sense of "normalcy" in the OIF years know of the difficulties in an effort to rebound after the aftermath of sanctions.

Before the deployment of forces to Desert Shield, the objectives were defined and the endstate was determined (see Chapter 1). What makes it a model for other conflicts to emulate was that not only was the plan developed prior to the fighting, but it was followed without adjustment; the planning assumptions and conditions didn't require any changes midstream; an anomaly when it comes to war planning-war execution histrionics. [11]

As explained in Chapter 1, the U.S. exit strategy in Operation Desert Storm was a model of simplicity:

- Mass forces in Saudi Arabia;
- Kick the Iraqis out of Kuwait;
- Go home.

With all the pundits arguing for more; one can not dispute that the objectives were met and the postconflict and war termination strategy was spelled out from the beginning. While the objectives were met, one of the stated objectives of "restoring peace and stability to the Gulf" was perhaps too vague and left an opening for those who wanted to go after Saddam to use it as an excuse for the future issues with Iraq. In fact, if the United States had gone on to Baghdad in 1991, the United States and its allies (those that stayed on for that fight—which would have been few) would probably have been involved in a counterinsurgency (COIN) fight for the past few decades.

Secretary of State, General (ret.) Colin Powell opposed going to war in 1990 because he feared another Vietnam, and once the decision was made to attack Saddam's forces in Kuwait and in Iraq, Powell was certainly not alone in seeing the prospect of another Vietnam even after the Iraqi Army had been broken in Kuwait and was fleeing northward. Many thought that the haste with which the Bush administration terminated the war reflected a Vietnam-driven dread of involvement in postwar Iraq. Many also felt that the fear of getting sucked into an Arab quagmire drove the Bush administration to end the war prematurely

With the Iraqi Army on the run, the administration took the extraordinary steps of declaring a unilateral cease fire in the absence of any request for terms from Baghdad. (See Chapter 1's discussion of war termination categories—category number 4 in this case talks about one side declaring unilateral victory. Desert Storm is an example of this termination strategy.) It then sent the Coalition Commander, General Norman Schwarzkopf, without political instructions, to Safwan, a place in allied-occupied Iraqi territory that was not even under U.S. control at the time of the cease-fire, to negotiate cease-fire terms with some of Saddam's military lackeys. What didn't appear obvious to the decision makers at this point was that the Iraqis should have been summoned to appear at Schwarzkopf's headquarters and told that a cease-fire required a public acknowledgment of defeat by Saddam himself.[12]

To many, the 1991 Persian Gulf War, while extremely successful, is a prime example of allied failure to take a holistic view of battle and develop a hierarchical set of objectives and associated measurable criteria to include the second portion of our endstate algorithm: conflict resolution. More importantly, an explicit set of objectives, many would argue, were not developed prior to the onset of hostilities to address what happens after war termination no matter how successful it was.

Arguably there has been a set of strategic-level objectives stemming from the 1990 "aggression will not stand" statement by President George Bush. Additionally, if such a hierarchy existed, the questions of whether the allied objectives were met and when the war should end would have been answered.[13] Regardless, the Administration's short-term goal for OIF was regime removal. As President Bush stated in his March 17, 2003, Address to the Nation, "It is too late for Saddam Hussein to remain in power." In that speech, he promised Iraqis, "We will tear down the apparatus of terror . . . the tyrant will soon be gone."[14] So there was obviously an endstate established before hostilities began.

President Bush declared in his television announcement of Operation Desert Storm on January 16 that U.S. objectives were not to destroy or to occupy Iraq. Implicit in Bush's announcement was that the driving of Saddam's forces from Iraqi would not necessarily result in the demise of his regime. But it was clear as early as the first day of the war that it would be difficult to draw a line between eviction of Iraq's occupational army from Kuwait, destruction of the remainder of its offensive and defensive military power, and removal of its regime from effective political and military control. The limitation of U.S. political objectives, as explained by President Bush, had more to do with keeping in line a precarious coalition of 29 countries, including key Arab states such as Egypt and Syria, than it did with a conceptual framework for the application of U.S. military power in restrained doses. Given Iraq's military weakness relative to the coalition mobilized against it, the agreed battle terms would be imposed, not negotiated, once war had started.[15]

In the end, Operation Desert Storm was more of an interest-based war and lent itself more to negotiation, persuasion, and coercion in the termination phase.[16] Typically even if a set of goals and objectives are developed beforehand, if there are resources involved like the magnitude involved in the Persian Gulf, then to reach a settlement requires a negotiated endstate. Even the ultimate outcome of war is not always to be regarded as final. The defeated state often considers the outcomes merely as a transitory evil, for which a remedy may still be found in political conditions at some later date.

In reality, the eight months of Operation Desert Shield, the lead up to the air and ground war in the early part of 1991, was a great benefit for it gave time to the planners and leaders in their effort to devise the postconflict goals. Although many argued there was no set of coordinated goals and objectives to achieve the national endstate prior to the commitment of forces, one could argue that the existence of Desert Shield or the buildup of ground combat power prior to Desert Storm helped the National Command Authority (NCA) develop a strategy. This grace period allowed the NCA to identify nested objectives (see Figure 6.1 and Chapter 7) and perhaps should be considered, in some respects, for future conflicts. Although the American military exited the war in the Gulf, it was, perhaps, a politically inconclusive departure that led to sanctions—a conflict resolution decision—and the next go-around for the United States in that theater: 2003.

GLOBAL WAR ON TERRORISM AND THE REEMERGENCE OF COIN THEOREMS

The reader will notice that in this text the Global War on Terrorism (GWOT) is both a "good" and "bad" example of war termination and conflict resolution use when defining an endstate. In short, there are significant successes and failures as they relate to GWOT. An ancillary good that comes out of the GWOT was the reemergence of the COIN doctrine. Although it took a number of years for COIN to be codified and, more importantly, resourced, the deployments to both Iraq and Afghanistan established a need for a structured counterinsurgency doctrine and a trained force to employ the complexity of its components.

Steve Biddle from the Center of Foreign Affairs asks the right question; "How would we know when the war has been won?" Unlike World War II or Operation Desert Storm, many wars like Iraq and Afghanistan will not end at an appointed hour by the signing of a peace agreement or the declaration of a cease fire. But any of these can have a discernable ending. For instance, and I will explain this later in greater detail, our desired endstate for the enemy threat in Iraq was the isolation of a remnant of al Qaeda into a small band of harried individuals living in deep cover as fugitives from the law, cut off from any base of popular support, despairing of any real hope of establishing their views through political power, and with no successor organization waiting in the wings to take up their struggle on behalf of a sympathetic people.

Defeating terrorism is the supposed endstate for the GWOT and is perhaps an insurmountable objective and arguably unachievable. Orthodox terrorism is not an existential threat to America; mass casualty terrorism on the scale of 9/11 and the ability to sustain this, by contrast, is not achievable by bands of isolated individuals. The record to date suggests that this magnitude of terror requires a degree of organization and profound commitment characteristic only of an institution like al Qaeda, and it is within our power to defeat al Qaeda as an institution even if we cannot kill every individual terrorist in the world.

The effort to protect the American Homeland by extending the security measures for travel and public gatherings coupled with the engagements abroad have arguably crippled the large terrorist networks' capacity to project power in the form of terrorism. Therefore, the activities in the form of combat operations against large terrorist networks have largely left them nullified and incapable of organizing like they had in the early stages of the GWOT. Although the arrival of this point may not be readily apparent, over time it will become ever clearer as the absence of mass fatality attacks on Americans grows prolonged.

Just as the Cold War's end was clear mainly in retrospect, we can expect that the end of this war will be proclaimed by historians rather than by soldiers. Looking backward today, we can say that the fall of Communist Poland, the destruction of the Berlin Wall, and the breakup of the Soviet Union signaled a period within which the Cold War ended, though no single event can be said to have provided more than a symbolic finale. Likewise, looking forward from today, there will come a time when we can be confident that we have seen the end of al Qaeda, but we cannot expect to be able to proclaim it at any single moment.

End, however, it shall—if we formulate our aims and our strategies properly.[17]

The objectives of al Qaeda in Iraq (AQI) during 2005 to 2006 were to drive the coalition from Iraq, undermine the government of Iraq (GoI) and ultimately establish an Islamic Caliphate. Following Abu Musab al-Zarqawi's public declaration of all-out war against the Shi'a in September 2005, AQI's statements and actions stoked sectarian violence by exploiting Sunni fears of Shi'a dominance. AQI claimed to be the bulwark against the Shi'a, and secured some Sunni support, particularly in Anbar. Anbar remained an AQI stronghold throughout 2005 and 2006, a platform for kinetic attacks into the capital, and a center for recruitment—and coercion—of the Sunni populace. Additionally, AQI violently discouraged Sunni participation in the government by targeting, publicly condemning, and physically attacking Sunni GoI members.

On the other side of the sectarian divide, the Shi'a, long repressed under Saddam, sought to maintain their new-found dominance. Jaysh al-Mahdi (JAM) was the most prominent and aggressive Shi'a group threatening Sunni survival and security. This illegal militia under the control of Shi'a cleric Muqtada al-Sadr drew on the young and largely uneducated population of the Sadr City district of Baghdad and other poor urban areas across the Shi'a areas of Iraq. Its goal was to protect the Shi'a and become a dominant force in Iraq by marginalizing the Sunnis. By late 2006, the corpses of dozens of executed Sunnis littered the streets of Baghdad and other Iraqi cities every morning.

During the summer of 2007, the war in Iraq saw an infusion of what was termed at the time Concerned Local Citizens (CLC) and then eventually the Sons of Iraq (SOI)—the establishment of security in the form of neighborhood watch tribal checkpoints. The inclusion of the Sunni majority in the form of population control was indeed a shift in how the war was progressing but perhaps was the genesis behind the beginning of the end of the struggle to defeat terrorism on a large scale.

So while I categorize GWOT in both the "good" and "bad" bins, we cannot ignore the fact that most of the effort thus far in the fight, especially in Iraq, is because of the incredible leadership of the likes of Odierno and Petraeus who stood firm on fixing the fight where post-hostilities perhaps was neglected to a degree at the onset of hostilities. Coming in after the fact to find a war termination strategy and, even more so, a conflict resolution vision is not easy as we have seen during the Surge of 2007 to 2008; let's hope our strategists can learn from this.

IRAQI FREEDOM

Irregular Warfare is far more intellectual than a bayonet charge.
                                                                    —T.E. Lawrence

Although Desert Storm may be, to an extent, the complete model for war termination and conflict resolution strategies, it is instructive to explore its impact on the follow-on Iraqi foray, Operation Iraqi Freedom (OIF). Even though the termination of war in Desert Storm arguably met the national objectives or endstates, the need for a conflict resolution was readily apparent in Iraq following the Gulf War, the imposed sanctions, and the eventual overthrow of Saddam Hussein.

In his March 2003 speech, President Bush declared that in the longer term, the United States would help Iraqis build "a new Iraq that is prosperous and free." It would be an Iraq, as he described it, that would not be at war with its neighbors, and that would not abuse its own citizens.[18] Those were the basic "endstate" elements typically used by war planners. The U.S. Central Command (CENTCOM) OIF campaign plan, for example, described the strategic objective this way: "A stable Iraq, with its territorial integrity intact and a broad-based government that renounces WMD [weapons of mass destruction] development and use and no longer supports terrorism or threatens its neighbors."[19]

Over time, since the days of war planning to the present, the Administration's longer-term strategic objectives have been fine-tuned. In the November 2005 *National Strategy for Victory in Iraq*, the Administration stated the long-term goal for Iraq: "Iraq is peaceful, united, stable, and secure, well-integrated into the international community, and a full partner in the global war on terrorism." In January 2007 at the time the "surge" was announced, the White House released an unclassified version of the results of its late 2006 internal review of Iraq policy. That document states: "Our strategic goal in Iraq remains the same: a unified democratic federal Iraq that can govern itself, defend itself and sustain itself, and is an ally in the war on terror." And in March 2008 in its regular quarterly update to the Congress, the Department of Defense used the same language almost verbatim: "The strategic goal of the United States in Iraq remains a unified, democratic and federal Iraq that can govern, defend, and sustain itself and is an ally in the war on terror."

Obviously, there existed a baseline framework that allowed the tactical and operational successes enjoyed in the first Gulf War. At the onset, OIF appeared to be a successful operation as well until it lapsed into an insurgency that few predicted; after four years of developing a counterinsurgency strategy, a "victory" of sorts appeared on the horizon. The development of a credible Iraqi Security Force (ISF) was actually the plan all along. However, the change in the type of conflict brought on by a set of poor conditions made by leaders in the theater had a major impact on the war termination and postconflict strategy.

Much of the success in OIF is based on the commanders, General David Petraeus and General Ray Odierno, recognizing a need to inject baseline objectives at the beginning of the 2007 so-called "surge" to feed into overall endstates defined by the military and fed to their political masters. What they did was initially to provide the clear-hold-build concept for countering an insurgency and then provide the resources to permit the success of the strategy. To allow for the success of this strategy they relied heavily on subordinate commanders in their "read" of their own battlespace and operation environment because they knew each part of Iraq was different from any other. Therefore, as I will briefly show here, much of what was achieved eventually in OIF was "bottom up" as opposed to the top-down Clausewitzian model for deriving endstate objectives.

The dramatic success in Anbar pointed the way for a Sunni anti-AQI Surge—a movement that became known as the Awakening or the Sons of Iraq (SoI). The SoI program remains a priority for the GoI, as evidenced by its uninterrupted funding in the face of fiscal constraint, and is arguably better-funded than any other program. The history of this reconciliation effort remains to be written but it is one of courage. Courage by the Coalition to reconcile with former insurgents, courage by those insurgents to cease fighting and seek to preserve Sunni rights through cooperation and political means, and courage by the mostly Shi'a elected leaders of Iraq to accept and support these former enemies and fellow Iraqis. It was a calculated risk; the only way this could have worked was to decentralize decisions to brigade, battalion, and company commanders. Senior commanders had to underwrite this risk and provide the top cover for their junior commanders. This is the type of dynamic leadership provided by the likes of the Corps and then Force commander, Ray Odierno, in resolving a conflict with no predetermined war termination or conflict resolution strategies.

TACTICAL SURGE OF IDEAS

Beyond the arrival of new forces, the most well known aspect of the Surge was at the tactical level: how the coalition embraced the clear-hold-build strategy. In past attempts, the Coalition had cleared, but Iraqi forces lacked the numbers and capacity to hold and build, allowing insurgents to reassert control of cleared areas through intimidation, extortion, and targeted violence on the ISF and civilians who cooperated with the Coalition or GoI. Coalition Forces, having returned to main bases, would be powerless to prevent the insurgents from reasserting themselves in cleared areas. The United States broke this cycle by moving forces out of the large forward operating bases (FOBs) and establishing them throughout communities in joint security stations (JSSs), patrol bases (PBs) and combat outposts (COPs), in units as small as platoons of 40 soldiers.

The combination of ISF and U.S. Forces together significantly increased our capacity to hold areas. The JSS served as a command and control node for all the Iraqi and U.S. Forces in an area, bringing together Coalition forces, Iraqi army, and the burgeoning police forces. JSSs also later served as neutral ground for grass-roots political meetings between tribal leaders, helping resolve disputes and serving as a forum for discussions. The Multi-National Force-Iraq (MNF-I) operated many of these stations side-by-side with Iraqi army and police and local volunteers, moving out into the communities, establishing relationships that would build trust. The concept was simple: protect the population where they worked and slept and their children played.

As General Petraeus frequently said, we were no longer commuting to the fight. Before the Coalition lived with the population, Iraqis who cooperated with the United States became the targets of retaliation attacks when Coalition Forces returned to our FOBs, attacks the ISF lacked the numbers and capacity to counter. Once established, the locals saw the troops were staying and relationships were built with key leaders and the general population. In fact, they often experienced the phenomenon that when they first entered an area, the population would be standoffish to U.S. soldiers and Marines until they saw the towering concrete barriers known as T-walls go up that indicated our presence would be semi-permanent. They would then immediately reach out to the force with information on the insurgency. They felt they would now be protected. The populace helped uncover and neutralize the militant groups threatening their homes and safety. Equally important, our presence working with partnered ISF units prevented the Sunni and Shi'a extremists from reasserting their presence.

Coalition Forces were only half of the clear-hold-build effort. Odierno and MNF-I recognized that we had to train and partner with the ISF for sustainable security and through this increase their capacity to fill the gap. The physical surge of troops enabled increased partnership with ISF at levels down to company formations. This new degree of partnering improved ISF capabilities exponentially compared to what military training and transition teams (MTT) alone had been able to do.

CLAUSEWITZIAN TRINITY

As a refresher, Clausewitz wrote that war is a phenomenon that, depending on conditions, can actually take on radically different forms. The basic sources of changes in those conditions lie in the elements of his "trinity." In simplistic terms, the *trinity* consists of the "the people, the army, and the government." According to Clausewitz, his trinity is really made up of three categories of forces: irrational forces (violent emotion, i.e., "primordial violence, hatred, and enmity"); nonrational forces (i.e., forces not the product of human thought or intent, such as "friction" and "the play of chance and probability"); and rationality (war's subordination to reason, "as an instrument of policy"). Clausewitz then *connects* each of those forces "mainly" to one of three sets of human actors: the people, the army, and the government.

1. The people are paired mainly with irrational forces—the emotions of primordial violence, hatred, and enmity (or, by implication, the lack thereof—clearly, it is quite possible to fight and even win wars about which one's people don't give a damn, especially if that is the case on both sides).
2. The army (which refers, of course, to military forces in general) and its commander are paired mainly with the nonrational forces of friction, chance, and probability. Fighting organizations deal with those factors under the creative guidance of the commander (and creativity depends on something more than mere rationality, including, hopefully, the divine spark of talent or genius).
3. The government is paired mainly with the rational force of calculation—policy is, ideally, driven by reason. This corresponds to the famous argument that "war is an instrument of policy." Clausewitz knew perfectly well, however, that this ideal of rational policy is not always met: "That [policy] can err, subserve the ambitions, private interests, and vanity of those in power is neither here nor there. . . . here we can only treat policy as representative of all interests of the community."[20]

Few would argue that there weren't challenges early on in the war in Iraq. The Civilian Provincial Authority under the leadership of L. Paul Bremer just didn't provide the coverage for policies involving the establishment of Iraqi Security Forces; local, provincial; and national governance; and economic programs to lead the newly liberated country to success. Instead, it relied on the military commands to establish the direction.

In fact, President Bush's major fault in his military strategy was nearly total disregard for the need for harmony in the Clausewitzian trinity. He asked for no shared sacrifices and put the entire burden for the wars on the less than 1 percent of the population that serves in the military. Congressman David Obey's November 19, 2009, idea of a war surtax perhaps would be a step toward curing that failing. Though a tax may be economically and politically unwise at the time, encouraging voluntary efforts by civilians (war bonds are examples from World Wars I and II) would be important symbolic steps toward sharing the sacrifice and moving closer to harmony in the trinity.[21]

Essentially, following the Clausewitzian models, wars and their end-states are theoretically defined by the highest level policy makers and executed by the military. The third part of the trinity, the public, is an extension of the policy being executed by the military. In OIF, the reverse essentially was the case. The tactical successes localized at the battalion and brigade levels were a forcing function to derive an eventual overall postconflict strategy. The development of *FM 3-24: Counterinsurgency Operations* in 2007 brought to light the need for a doctrine-driven strategy. As a result, the Coalition leaders in Iraq established a universally followed strategy that led to multiple victories. The first few years of OIF saw achievements of tactical objectives that built on an overall strategic exit strategy that gives the reader a sense of how soldiers' actions forced the hand of the policy makers. I argue that this bottom-up process, while appearing to be effective at this point, comes at a cost.

After the 2003 to 2004 debacle that involved disbanding the Iraqi Army, only to restart it from scratch, coupled with the disenfranchised population brought on by the Administration's ignorance of the power base in a Muslim secular society, the Coalition stumbled into the next phase. In this case, what was next was a form of stability operations. Unlike our inability to see the emergence of an insurgency after "the end of major combat operations"[22] occurred in the spring of 2003, this time, fortunately, it was the population recognizing and then choosing peace over violence (see boxed text, Clarifying and Updating U.S. Interests and Strategic Objectives in Operation Iraqi Freedom).

Unfortunately, the challenges brought on by the Coalition Provisional Authority by the 2003 disbanding the Iraqi Army and de-Baathification of the populace derailed that portion of the endstate. Reconstruction and nation-building truly were afterthoughts. This is a striking example that causes planners to put thought into how military operations and ensuing decisions to eliminate or degrade an adversary's national infrastructure, institutions, and organizations may affect the postoperations military and political objectives. Sanctions and destruction of power grids in the early stages of a fight are two examples that do just that. As discussed briefly, sanctions placed on Iraq in the 1990s led to an extreme degradation of the nation's infrastructure leaving the invading force responsible for reconstituting the nation so it could function properly—a difficult task at best.

After the execution of the bottom-up driven strategy or the Surge beginning in early 2007, Iraq has become a nascent democracy that is rebuilding its strategic depth as a regional power in the Middle East. Given its strategic location and national resources, Iraq remained a vital interest to the United States. Implemented in 2009, the U.S.-Iraq Security Agreement (governing the temporary presence of U.S. Forces through 2011) and the Strategic Framework Agreement (an open-ended partnership) reflect the maturing relationship between the two countries and will serve as the foundation for a comprehensive, long-term partnership.

Derived from the path set out by General Odierno once he took command of the Multinational Forces-Iraq in the summer of 2008, newly elected President Barack Obama outlined his goals in February 2009: an Iraq that is sovereign, stable, and self-reliant with an Iraqi government that is just, representative, and accountable and contributes to the peace and security of the region. On September 1, 2010, combat operations in Iraq officially ended and a transition force of about 50,000, including advisory

CLARIFYING AND UPDATING U.S. INTERESTS AND STRATEGIC OBJECTIVES IN OPERATION IRAQI FREEDOM

Announcing the drawdown and transition policy, President Obama stated that the goal is "an Iraq that is sovereign, stable, and self-reliant." To that end, the President added, the United States would:

- Work to promote an Iraqi government that is just, representative, and accountable and that provides neither support nor safe haven to terrorists; help Iraq build new ties of trade and commerce with the world; and

- Forge a partnership with the people and government of Iraq that contributes to the peace and security of the region.

Under that broad rubric, as the U.S. role in Iraq transitions, it might be useful to confirm key U.S. national interests regarding Iraq, and the crucial strategic objectives that, at a minimum, it is important for the United States to achieve to support those interests. Such broad objectives might address the following elements:

- U.S. interests in Iraq's domestic political arrangements. Some might argue that a democratic or broadly representative and inclusive Iraqi polity is essential as a key to Iraq's stability, while others might argue that the nature of Iraq's domestic political arrangements is much less important than simply a unified and stable Iraq.

- U.S. interests in Iraq's role in the fight against global terrorist networks. Some might argue that the most important goal is simply ensuring that Iraq does not serve as a safe haven for terrorists. Others might stress the importance of active intelligence-sharing by Iraq with the United States. Still others might argue that it is in U.S. interests that Iraq couples the counterterrorism skills it is currently developing as part of its domestic counterinsurgency effort, with expeditionary capabilities, so that it could participate in future regional counterterrorist activities.

- U.S. interests in the regional balance of power. Some might argue that Iraq's strength, relative to that of its neighbors, is important. Others might simply stress the importance of an absence of conflict—that is, as a long-stated U.S. goal puts it, an "Iraq at peace with its neighbors." Furthermore, it may prove judicious to update the formulation of U.S. strategic objectives as the U.S. mission and presence in Iraq change and results of those changes are assessed.

In his policy announcement, President Obama stressed that the situation in Iraq remains dynamic and challenging: But let there be no doubt—Iraq is not yet secure, and there will be difficult days ahead. Violence will continue to be a part of life in Iraq.

Source: "Operation Iraqi Freedom: Strategic Approaches, Results and Issues for Congress," Catherine Dale, Washington DC: CRS, April 2, 2005, p. 12.

and assistance brigades, special operations forces, and enablers were the residual part of what earlier in the year had become U.S. Forces-Iraq.

Capitalizing on the successes achieved by local commanders, the United States was then able to define a conflict resolution strategy. To establish strategic depth, Iraq must rebuild its institutions and infrastructure. Since the beginning of the war, the combined, sustained efforts of U.S., Coalition, and Iraqi Security Forces—coupled with the efforts of civilian partners—have reduced security incidents and attacks of all types to the lowest levels since 2003, allowing the Iraqi government to focus on capacity building. Although statistics do not paint the whole picture, they provide context to understanding the progress made to date. As a measurement in defining success as Iraq capitalizes on the capacity-building efforts instigated by the U.S.-led international effort, by the time of the March 2009 National Elections, overall security incidents had decreased 78 percent, civilian deaths 76 percent, and U.S. casualties (killed and injured) 87 percent.

Emerging from more than 30 years of authoritarian rule based on ethnosectarian privilege, Iraq is slowly addressing the myriad of challenges to its future as a stable, multiethnic, representative state. The March 2010 elections set the stage for the next four years of progress. Iraq is a state and a society under construction, struggling to define its identity and place in the world after decades of oppression and violence. The goal of any COIN campaign is to win over the proverbial hearts and minds of the populace and to convince them to back the government and oppose insurgents. This requires providing every citizen with basic services like electricity, food, and clean water; law and justice; security against arbitrary reprisals; and a functional economy in which the people are able to support themselves and their families. Of course, none of this is possible without reasonably good governance to ensure that resources are being properly allocated as well as procedures properly developed and applied to ensure the security and welfare of the people.

The U.S. military presence through 2011 provides psychological and physical support to the Iraqi people. It will allow them to build political constituencies and alliances, build an economic foundation, and continue to develop the Iraqi Security Forces to assume all security responsibilities by December 2011. The level and nature of U.S. engagement with the Iraqis will continue to change as the U.S. military draws down and transitions responsibilities and activities to the U.S. Embassy in Baghdad, the Iraqi Government and Security Forces, and other nongovernmental organizations. Iraq is making steady progress, but it still has a long way to go. The United States must have strategic patience. Through the Strategic Framework Agreement, the United States has a mechanism for supporting Iraq to develop its institutional and human capacity.[23] OIF continues on as of this writing. Victory appears on the horizon and although the endstate conflict resolution strategy wasn't immaculately planned, there was validity in the original plan, however stove-piped it appeared.

Perhaps in remembrance of the confusion and shifting enemies faced so recently, the Coalition Forces at higher headquarter levels display surprising reluctance to accept the full extent of stability and revitalization in Iraq. Though Iraq may never look like a Western-style democracy (and perhaps Iraq *should* never look like a Western-style democracy), and many years still lie between the current Iraqi government and a truly stable and productive political society, the state is doing pretty well. Though we may have a hard time believing this, Iraqis, only a few years ago, lived just fine without us; and they will do so again and under the construct of government they eventually decide as a nation. True, the rift between the Shi'a authorities in Baghdad and the Sunni tribal lands surrounding the city runs deep; true, the infrastructure remains patchwork at best; true, the Army currently performs many of the functions that a civilian government should, forming, effectively, a military state in many parts of Iraq. True, also, that the Iraqi Army shows a degree of professionalism and competence that has, in many ways, progressed in the past few years, but there is still a long way to go to make it a formidable force.

The brigade combat team (BCT) that I commanded, 3rd BCT of the 101st Airborne Division came to the theater with a plan to transition to the ISF, which was its war termination strategy at the tactical level. I expound on my personal experiences later as we discuss the "Whole of" approach in Chapter 5 where I attempt to define an interagency strategy. In short, my premise is that if there had been a plan for "what's next?" then serendipitous transitions like that of high-intensity conflict to an immersion of insurgencies, on one end of the spectrum, and the turning of the population to peace on the other end would have been nested from the onset.

URUGUAY AND THE TUPAMARO

Uruguay in the 1960s was distinct among South American counties for its affluence and sociopolitical stability. Economic prosperity had fostered the growth of a large middle class and a stable welfare-state government that allowed a wider degree of democratic and civil freedoms than any other South American government. Because Uruguayan society was so peaceful, the army and police were very small. In 1968, there were only about 12,000 men in the armed forces and fewer than 22,000 police to keep order in a population of about 3 million.

After the Korean War, a slump in the demand for wool and meat, Uruguay's two principal exports, brought mass unemployment, inflation, and a steep drop in the standard of living. The social tensions this produced, along with the corruption of the overblown state bureaucracy (one in five working Uruguayans was employed by the federal government in some fashion), gave the impulse for an effective urban guerrilla movement to emerge.

This revolutionary group's official name was Movimiento de Liberación Nacional (MLN or National Liberation Movement), but was popularly known as the Tupamaros (from Tupac Amaru, last member of the Inca royal family, murdered by the Spanish in 1571). Their current political party is known as the MPP (*Movimiento de Participación Popular,* or Popular Participation Movement). At one time, there may have been as many as 5,000 Tupamaros operating in Uruguay. It was founded in 1963 by Raul Sendic, a law student studying in Montevideo. Because Uruguay was so urbanized, with more than 80 percent of Uruguayans living in large towns or cities, they concentrated almost all their activity in and around the capital, Montevideo, where more than half of the country's entire population lived. As with most other South American guerrilla groups, the Tupamaros started as a political organization that deliberately chose to be seen as an "earned struggle" drawing its membership from the young, radicals, and middle class—mostly students and white-collar workers.

Like most urban terrorist groups, they were organized in a cellular structure of four to five men with the group leader as the only link to other cells. This was done for security reasons, as was their practice of never telling any individual more than he or she needed to know for any particular operation. From 1963 to early 1968, the Tupamaros concentrated on gathering resources—mostly by robbing banks, gun shops, and private businesses. Their goal was to make the government look powerless to defend its populace and unnecessarily heavy-handed in its reactions. The main tactics they used to achieve this were political kidnapping and intimidating the security forces. The Tupamaros used political kidnapping as an alternative to assassinations and as a way to show the government's impotence. It was a severe psychological shock and embarrassment to those in the government to have their friends and diplomatic figures snatched off the street and held in so-called people's prisons in Montevideo itself. Meanwhile, the people did not react as they would have against assassinations because the kidnappings of unpopular and corrupt people did not inconvenience them, but the inept police reaction—usually a massive cordon-and-search operation—did.[24]

Beginning in 1968, the government resorted to ever more drastic repression to try to snuff out the insurgents and crush their allies among trade unions and other civil society groups. In 1971, the government transferred responsibility for the COIN campaign from the national police to the armed forces. The army responded by ratcheting up levels of domestic repression to previously unheard of heights to the point where Uruguay had the greatest number of political prisoners per capita in the world. Simultaneously, the government suspended its citizens' individual liberties to facilitate the military's counterterrorism and COIN measures.

When it was committed to full-scale COIN operations against the Tupamaros, the Uruguayan military realized that the critical factor sustaining the insurgents was popular support, especially from the frustrated middle class, as a result of the woeful state of the Uruguayan economy. The military also concluded that this economic malaise was the product of the corruption and mismanagement of successive Uruguayan governments. Consequently, in addition to its all-out war on the Tupamaros, the Uruguayan military waged a slightly less intense war against "economic crimes" committed by government officials. So the political and military endstate was to defeat an insurgency and enhance the economic capacity of the state. To do so, the Uruguayan military defined its war termination strategy as defeating the Tupamaros so that the insurgent group lost popular support and turning the populace to the side of the military. The conflict resolution strategy was to bolster the nation's economic base so that a democracy could flourish. Knowing both strategies helped the indigenous security force achieve its objectives.

As a result, the military increasingly arrogated political powers to itself. The worsening state of the economy provoked a wave of student rioting and labor unrest, and a state of national emergency was declared in June 1968 (which was to last until late 1972). It was during this crisis that the Tupamaros staged their first political kidnapping—Ulises Pereyra, the president of the State Telephone Company, an unpopular figure whose abduction was acclaimed by the public. When the police began to search the campus of the National University in Montevideo, they started a student riot that ended in the death of a student. Ulises Pereyra was released unharmed five days later. More kidnappings followed. In September 1969, they kidnapped a leading banker and held him for 10 weeks in support of a strike by employees at his bank. In July 1970, Dan Mitrione, an American policeman on loan to the Uruguayan security forces, and Aloisio Gonide, the Brazilian consul in Uruguay, were kidnapped and held for ransom. When the government refused to parley with the Tupamaros, they killed Mitrione, a move that was to cost them considerable public support. In the first half of 1971, the British ambassador to Uruguay, the Uruguayan attorney-general, and a former minister of agriculture were kidnapped, and Ulises Pereyra was abducted a second time.

In February 1972, President Juan Maria Bordaberry tried to check the military by naming a new defense minister he hoped would rein in the soldiers. The generals countered by demanding that he undertake a set of specific economic reforms, which they believed would undermine the internal unrest, and then forced Bordaberry to give up much of his power to a military-dominated national security council. Within a year, this junta had dissolved the parliament, closed the national university, banned virtually all Leftist parties, and imprisoned its political opponents.

As stated, a tactic used by the insurgents was to attempt to intimidate the security forces. Because the police were doing almost all of the fighting against the guerrillas, the Tupamaros began to select individual policemen for assassination in late 1969. Although only a few policemen were killed this way, the morale of the force was shaken. In June 1970, there was a general police strike for higher pay and the right to work in civilian clothes to make them less conspicuous. The government's response to the insurgents was uneven and predictably clumsy. Although at first the army and police were small and untrained in counterinsurgency techniques, a paramilitary body of 20,000 men called the Metropolitan Guards was set up in 1968. It was trained by American and Brazilian policemen, and its main duty was to oppose the Tupamaros. The security forces were handicapped by a lack of reliable informers, a coordinated network for sharing and disseminating intelligence, and their habit of conducting massive cordon-and-search operations that more often than not alienated the public's support. It appeared to many people that the Tupamaros, who by now had reached an active strength of almost 3,000, were running rings around the police. The economy failed to improve, and a series of corruption scandals in 1970 and 1971 further undermined public support for the government. It seemed that the Tupamaros were on the verge of creating the climate of collapse that would lead to the government's fall.

In November 1971, presidential elections were held. An alliance of left-wing parties called the Frente Amplio emerged to challenge the rule of the two traditional political parties. The Tupamaros gave vocal support to the Frente Amplio while continuing their campaign of provocation. However, the murder of Dan Mitrione in August 1970 and the continual civic disruption pushed more and more people away from support of the Tupamaros. The Frente Amplio suffered from the association and got less than 20 percent of the votes in the elections. The new Uruguayan president, Juan Maria Bordaberry, suspended civil liberties and declared a state of internal war with the Tupamaros in April 1972. The Army, which until this point had been playing only a supporting role to the police and Metropolitan Guards in the counterinsurgency, was ordered into action. They relied on mass arrests, torture, and large cordon-and-search operations. These saturation tactics captured most of the guerrillas and forced the remainder to flee the country. By November 1972, the Tupamaros had ceased to be a threat to Uruguay.

The government had won, but only at the cost of destroying democracy in Uruguay and alienating large sections of the population. The army, which in 10 years had gone from consuming 1 percent of the national budget to more than 26 percent, was not about to go meekly back to the barracks. In their view, they had just cleared up the politicians' mess and were not going to let it happen again. The military leadership pressured President Bordaberry to maintain the declared state of internal war. By mid-1973, all left-wing political activity had been suppressed and the national legislature indefinitely dissolved. Uruguay, once the most tolerant and democratic country in South America, had become another garrison state. This was the only permanent legacy of the Tupamaros, although they had advanced further and offered a more serious challenge to established power than any other urban guerrilla movement.

In the mid-1980s, democracy returned to Uruguay and the Tupamaro movement went legit, laying down their weapons in favor of joining the political process. By 1984, the Uruguayan people had seen enough of the military government. They took to the streets, demanding democracy. President Gregorio Alvarez organized a transition to democracy, and in 1985 they held free elections. Julio María Sanguinetti of the Colorado Party won and immediately set about rebuilding the nation. As far as the political unrest of the previous years, Sanguinetti settled on a peaceful solution: an amnesty that would cover both the military leaders that had inflicted atrocities on the people in the name of counterinsurgency and the Tupamaros who had fought them. The military leaders were allowed to live out their lives with no fear of prosecution and the Tupamaros were set free.

LEADERSHIP ALWAYS MAKES A DIFFERENCE

In *A Question of Command: Counterinsurgency from the Civil War to Iraq*, Mark Moyar believes that even sound counterinsurgency methods "have consistently failed when good leaders were lacking." Good leaders, however, can independently derive the principles of population-centric counterinsurgency or any postconflict challenge. Moyar tests his leader-centric COIN theory against nine cases—the American Civil War, Reconstruction, the Philippine Insurrection, the Huk Rebellion, the Malayan Emergency, the Vietnam War, the Salvadoran Insurgency, and the current wars in Afghanistan and Iraq—and discovers that flexible, creative, well-organized leaders played critical roles in achieving successes in these wars.

Poor leaders, perhaps unsurprisingly, fail to defeat insurgencies. The solution is clear to Moyar: Rather than providing doctrine based on historical experience on how to conduct counterinsurgency successfully, military institutions should choose good leaders and let them figure it out. For Moyar's argument to hold, he would have to demonstrate that good leaders can succeed using enemy-centric counterinsurgency tactics and that leadership, rather than adherence to the classic principles of counterinsurgency, determines success or failure in this kind of war. Unfortunately for his thesis, that is not the case; instead, in every instance, creative, charismatic commanders developed the same methods of conducting counterinsurgency focusing on intelligence, coordinating civil and military authority, working to gain the support of the population, building capable local self-defense forces, and protecting the population by stationing forces among them. And in every case, the demands of countering insurgency in a given situation "presented variations and changes that demanded frequent adaptation."

These are not revolutionary conclusions; the Introduction to *FM 3-24* states that "by focusing on efforts to secure the safety and support of the local populace, and through a concerted effort to truly function as learning organizations, the Army and Marine Corps can defeat their insurgent enemies."[25]

Iraq is a prime example for in 2006 it was a tangled web of groups with differing agendas, needs, fears, and capabilities, interacting in complex and unpredictable ways. In that way, it was like every other conflict. The surge of ideas brought on by the counterinsurgency mindset contained elements that could work in other operations, but the dilemmas of every war mean that no idea's effectiveness can be predicted with certainty. Each operational environment is unique, demanding unique solutions. Nonetheless, many of our successful initiatives could be useful in other theaters, and in preparing the U.S. army for the future.

The leveraging of the government across the interagency, focusing the network of our national agencies, each bringing their unique strengths and competencies to the immediate fight, is necessary for future success. Unity of effort through unified action was the key to this interagency effort. The physical form and location of that cooperation will depend on individual circumstances, but the concept of fighting a network with a network is a principle that the military and the larger U.S. government should institutionalize. Leveraging technology by moving the information created by that network as close as possible to the battlefield is the inseparable next step. These initiatives build on the permeation of information technology through our society and will often be a major asymmetrical advantage for U.S. Forces. When coupled with a surge in intelligence assets to the lowest possible tactical level, these advances permit agile operations with unmatched precision—a requirement on today's COIN battlefield.

Other parts of our Surge strategy hold potential, but require more careful consideration. Clear-hold-build as a strategy is dependent on many variables, not least of which are the availability of quality troops and the density with which they can hold key terrain. Tactical reconciliation is an excellent concept, but its implementation depended on the characteristic structure of Iraqi society and AQI's threat to its traditional power structure. Partnering with local forces is similarly thorny. Although essential to COIN success, the forces may not be good enough to make a difference.

One advantage the United States had was Iraq's history of a national army rooted in the colonial period. Not all theaters will have such a strong base on which to build a modern force and secure the state. At the operational level, the indirect approach brought us success. To paraphrase Hart, it loosened AQI's hold by upsetting their balance. Like many strategic ideas, the concept may be clear, but execution is more challenging. The key difficulty is identifying the enemy's vulnerability, that is, how to push them off balance. The seminal breakthrough was capturing AQI's operational scheme, which identified the areas surrounding the capital of Baghdad (known as the Belts) as key terrain. The hard work of our operators and intelligence analysts may have been essential, but so was good fortune. Finally, the effectiveness of the tools used to build civil capacity—provincial reconstruction teams, embedded provincial reconstruction teams, Commander's Emergency Response Program, our bottom-up approach—are highly dependent on the structure of the society we're operating in, the strength of its government, and the links between the two. These are complex and difficult to discern without close and continuing contact with the society.

All of these innovations, though, depend on the one thing that is guaranteed to transfer to any theater—the thought process of shaping and adapting to the environment. Blindly following any example or doctrine is dangerous. It is also foolish, because the ultimate reason for our success in the Surge is guaranteed to be present wherever the U.S. Armed Forces fight: the intelligence, adaptability, courage, and strength of the U.S. soldiers, sailors, airmen, marines, and coast guardsmen. We have developed a generation of leaders who have tremendous tactical experience in some of the toughest terrain imaginable against a tenacious enemy. Given the freedom to apply their hard-won skills in the framework of a strategy that marshals all the resources of the nation behind them, they will bring victory.

NOT JUST NICE TO HAVE

Although it is obviously beneficial to have a war termination and conflict resolution strategy planned in advance, it is also a *requirement* in accordance with international law. Most conflicts require military occupation that places the military responsible under a number of international conventions for the treatment and care of the civilians in the occupied territory.

Regulations attached to the 1907 Hague Convention stipulate that the occupier is obliged to take all measures to restore and maintain public order and public life. The Fourth Geneva Convention of 1949 protects populations against the effects of war—in particular the protection of inhabitants of occupied areas. The First Additional Protocol of 1977 to the Geneva Conventions indicates that the intended military advantage of an action must constantly be weighed against the resulting disadvantage to the civilian population. This obligation also applies to the occupying force.[26]

The Geneva Conventions and their Additional Protocols are at the core of international humanitarian law, the body of international law that regulates the conduct of armed conflict and seeks to limit its effects. They specifically protect people who are not taking part in the hostilities (civilians, health workers, and aid workers) and those who are no longer participating in the hostilities, such as wounded, sick, and shipwrecked soldiers and prisoners of war. The Conventions and their Protocols call for measures to be taken to prevent or put an end to all breaches. They contain stringent rules to deal with what are known as "grave breaches." Those responsible for grave breaches must be sought, tried, or extradited, whatever their nationality.

The Geneva Conventions, which were adopted before 1949, were concerned with combatants only, not with civilians. The events of World War II showed the disastrous consequences of the absence of a convention for the protection of civilians in wartime. The Convention adopted in 1949 takes into account the experiences of World War II. It is composed of 159 articles. It contains a short section concerning the general protection of populations against certain consequences of war, without addressing the conduct of hostilities, as such, which was later examined in the Additional Protocols of 1977.

As stated, the bulk of the Convention deals with the status and treatment of protected persons, distinguishing between the situation of foreigners on the territory of one of the parties to the conflict and that of civilians in occupied territory. It spells out the obligations of the Occupying Power vis-à-vis the civilian population and contains detailed provisions on humanitarian relief for populations in occupied territory. It also contains a specific regime for the treatment of civilian internees. It has three annexes containing a model agreement on hospital and safety zones and model regulations on humanitarian relief. The addition of these articles to the rule of international law as it relates to conflict establishes a framework from which war termination and conflict resolution planners can begin to detail the vision for establishing a peaceful and functional environment for a war-torn nation.

CHAPTER 4

The Bad

War involves a train of unforeseen and unsupported circumstances that no human wisdom could calculate to end.

—Thomas Paine, British pamphleteer, revolutionary, radical, inventor, and intellectual

So what happens when things don't go as planned or go wrong because of a lack of planning? Regardless of the dynamics of conflict brought on by the likes of an insurgency or other campaign-changing events, an exit strategy is highly desirable because lacking it could result in reduced confidence in leadership, a drop in troop morale, possibility of increased casualties, and may negate any success achieved by the actual intervention which, in turn, may negate public opinion and support.[1] But these detriments hit only on the operational effects; strategically, the cost could be cataclysmic.

As seen in Operation Iraqi Freedom (OIF), the slow transition to a manageable conflict resolution strategy may result in the onset of an insurgency because the environment became vulnerable to insurgent activity. Examples of failures or conflicts we can learn from include the Korean War, 1956 Suez Crisis, Vietnam, Somalia, Bosnia, the Global War on Terror, and Operation Enduring Freedom (OEF).

KOREA . . . THE NEVER-ENDING (TERMINATING) WAR

Numerous theories have been put forward, even in this book, to explain why a state decides to terminate a war. This portion of the book will examine the following four theories as they relate to the Korean Conflict; a conflict that by my estimate was an example as a failure in war termination and conflict resolution:

1. Winners and Losers: This theory states that a war ends when one state wins and the other state is defeated. The theory predicts that when a state's forces

are decisively defeated and the state's leaders realize that they have lost the war, they will be compelled to seek an end to the war. This is applicable even if the loser's forces are exhausted through attrition warfare and there is a time lag between the turning point in battle and the leaders' decision to quit fighting.

2. Cost/Benefit: This theory assumes that the decision to terminate a war is a rational one, involving cost-benefit calculation. A state will pursue its objectives or war aims until it reaches a point where the marginal costs of continuing the war are not worth the objective. The state's leader will then decide to seek a termination to the war. This theory emphasizes the need for states and their leaders to act as if they are evaluating their actions against their objectives and to continually re-evaluate their circumstances in the light of new evidence.

3. Hawk and Dove: This theory by Fred Iklé, author of *Every War Must End,* states that statesmen who lead their country into war are too committed to that course to be able to rationally calculate or even to change their minds. Instead, they are replaced by new leaders who are less committed to the war and who seek peace. Perhaps the US Administration transition from Bush to Obama is an example of this theory.

4. Second Order Change: This theory by Joseph Engelbrecht states that when leaders are faced with a dramatic political decision such as war termination, they undergo a psychological process that forces them to see the problem from a higher or second-order paradigm. This theory argues that leaders involved in the absorbing commitment of war concentrate on the task at hand, the means or solution to their aims. Their very persistence makes it difficult to consider other courses or judge incoming information. At some point the leaders realize the attempted solution a la war becomes the problem. From a second-order perspective, leaders recognize that the war threatens another, perhaps a higher value which they hold dear. Thus, they see war termination as a necessary part of a future policy aimed at protecting this value.[2]

In the case of the Korean conflict, the events leading to the signing of the armistice and whether the four theories can adequately explain the "termination" of the Korean War are explained below. Note that this is merely an instructive way of showing why nations begin a fight and the theories behind how they work to either maintain that fight or work to terminate it. These theorems by no means replace the premises offered in this work; instead the intent of presenting each is an attempt to augment the overall concept of this work.

Winner/Loser Theory

When the Korean War ended, there were no winners, only losers. The Korean people were the biggest losers. Many Koreans had to evacuate their homes. The number of civilian casualties was estimated to be more than the armies of the Communists and the UN forces combined. North Korea's aim to reunite the whole Korean peninsula was not achieved.

Instead, it actually lost more real estate as the final signing of the armistice put the Demilitarized Zone even more to the north of the 38th parallel. The communists actually wanted the armistice more than their foes on the UN side. In fact, it not been for Stalin, the war arguably would have ended in 1951.

For the UN, the initial aim was to drive back the communists over the 38th parallel and reestablish the boundary between North and South Korea. But following the successful landing of the UN forces at Inchon, the UN decided to fight for the reunification of the Korean Peninsula under a democratically elected government as its objective. This aim was not achieved at the signing of the armistice in July 1953.

There is no decisive winner in the Korean War. Both the communists and the UN forces were in favor of the armistice. The only exception was the South Korean President, Syngman Rhee, but by then, President Rhee was not really in control of the war. Therefore, in short, the Winner/Loser theory cannot adequately explain the termination of the Korean War.

Cost/Benefit Theory

As early as 1951, the Chinese and the North Koreans suggested to Stalin to end the fighting as both of them had sustained heavy casualties in the war. In a letter from Mao Zedong to Stalin on July 20, 1951, Mao presented the following analysis of the Korean War in which his interest in achieving peace in Korea was obvious:

> . . . We . . . should revise our stand on foreign troops in Korea. Earlier, for the sake of restoration of two separate North and South Koreas, we agreed to stop hostilities at the 38th parallel. We agreed to this mainly because our military forces are capable of chasing the enemy out of North Korea but they are not strong enough to chase the enemy out of South Korea. In the event of a prolonged war, we may inflict even more damage upon the adversary, but we'll severely undermine our finances, it will be difficult to increase our defense construction. If we continue to fight for 6–8 months, we might be able to chase the adversary out of South Korea, but the price will be too high, we'll face a crisis . . . if the war . . . resumes again, we'll have to continue a long war trying to achieve the unachievable . . .[3]

In 1951, 60 percent of China's tax revenue went to the defense budget. The military weapons from Russia were not free. By the end of 1952, the losses were more than 1 million Chinese soldiers. It must be remembered that China had just ended a civil war on their homeland and was looking to reunification with Taiwan. The Chinese did not favor the war in Korea, but were pressured by Stalin to assist the North Koreans. China suffered a severe drain of experienced manpower, and as evident in his letter to Stalin, Mao was of the opinion that there was little benefit in prolonging the war as the cost of the war was too high to justify any benefit that could be gained.

For North Korea, the loss was more than 520,000 soldiers and more than 1 million civilian casualties. There was widespread shortage of food and the people were demanding an end to the war. In a letter to Stalin dated July 17, 1952, Kim Il Sung complained to Stalin: ". . . The adversary, almost without any losses, constantly inflicts upon us enormous losses in human lives and material valuable. . . ."[4]

By the end of 1952, a total of about 142,000 U.S. casualties was estimated (see Table 4.1). The U.S. casualties were estimated at more than 1,000 per week during the period of negotiations on the terms of the armistice. The United States had committed more than 1.2 million troops in Korea and more bombs had been dropped than had been used in World War II. Moreover, the American public and the soldiers in Korea were not sure if the prolonged war served any purpose and some questioned the need to send U.S. soldiers to fight in Korea. President Truman did not run for re-election and if he did run, he most likely would have lost the re-election due to adverse opinions at home arising from the Korean War.

When Eisenhower took over the presidency in 1953, he remarked ". . . U.S. is spending itself into bankruptcy," aptly summing up the view of the Americans on the Korean War. The benefit was not worth the additional real estate that could be gained from the prolonged war in Korea. The United States was very much in favor of the armistice. The ground war had been in a condition of stalemate since June 1951 and both sides were maintaining large armies along a static battle line at a cost exceeding the military value. Both sides had a simple choice: broaden the war to reach a decision or arrange a truce. The United States had opted for a "limited war" or risked losing the support of the other UN forces and also

Table 4.1   Casualties in the Korean War Conflict

Deaths	33,629
Killed in Action	23,300
Wounded in Action	105,785
Died of Wounds	2,501
Missing in Action	5,866
Died While Missing	5,127
Returned from Missing	715
Still Missing	24
Captured or Interned	7,140
Died While Captured.	2,701
Returned from Capture	4,418
Refused Depatriation	24
Total Casualties	142,091

risked the potential outbreak of World War III. Similarly, the Chinese had no desire to extend the war to other areas and risk air and naval bombardment of their mainland as recommended by General MacArthur.

In the end, both sides decided the cost was too great for both to carry on with the war. The only stumbling block was Stalin, but that, too, was removed in March 1953 when the Soviet leader passed away.

Hawk/Dove Theory

By June 1951, it was evident that the Chinese and the North Koreans had suffered heavy losses and were not keen to prolong the war. In June 1951, Kim II Sung and Gao Gang went to Moscow to convince Stalin to agree to the necessity of an armistice. While the Soviet leader gave his consent, he was in fact interested in prolonging the Korean War. In his assessment, the war tied the hands of the United States and stirred up tensions among the Western allies and within the United States as well. It also prevented the Chinese from befriending the United States, which had been one of Stalin's fears since the 1930s when Mao had confidential talks with U.S. representatives.

From the correspondences between Stalin and his allies, Stalin was constantly persuading the Chinese and the North Koreans to prolong the war. On November 19, 1951, Stalin sent a letter to Mao telling him to carry on negotiations for the armistice "without a strong interest in the early completion of negotiations." On March 7, 1952, he passed a resolution stating that the communists should not "hurry the process of negotiations" as it was not in their interest. Throughout 1952, Stalin constantly repeated to Mao and Kim to prolong the war as the United States needed it more than them and that the war had shown the weakness of the United States.

That Stalin was the "Hawk" in the Korean War is very evident as the Communists quickly reversed their position immediately after Stalin's death on March 3, 1953. The Kremlin was quick to write to Mao and Kim on March 19 calling for an end to the war and expressing the desire to achieve peace. The letter specified exactly what the Chinese and the Koreans were to do to terminate the conflict and to achieve peace. The Americans were prepared to avoid a breakdown in the negotiations and showed some flexibility. As General Bradley remarked, "It is better to make concessions than to risk an all-out war with China."[5] The failures of the initial negotiations were due mainly to pressures from Stalin to prolong the negotiations.

Second Order Change Theory

There is little evidence to show that this theory can be applied to explain the termination of the Korean War. Truman did not choose to end the war even though his political career was badly affected by the prolonged negotiations from June 1951 until he handed over the presidency to Eisenhower. In fact, Truman was conscious of the fact that although it was desirable to end the war quickly, it was not to be done at the expense of conceding too much advantage to the Communists. In 1952, he chose not to run for re-election.

For the Communists, Stalin's death changed nothing except to allow the "Dove" to end the war. The Communists were suffering from the loss of lives, but they were still firmly in power. The Chinese and the Russians were very much aware that the UN Forces were not willing to extend the war into China and Russia, so there was no threat to their national survival. At the same time, it was clear to the North Koreans that the Chinese and the Russians would never allow the UN Forces to take over North Korea. The only dramatic change during this period was the death of Stalin and the change in the U.S. presidency. Still, the result of this change had not resulted in any second-order change in the Korean War.

Of the four theories discussed, it is evident that the main reason for the armistice to proceed was the death of Stalin, who had been the stumbling block during the negotiations. Stalin died in March 1953, and the armistice was signed four months later in July.

The Winner/Loser Theory cannot be used to explain the termination of the Korean War as there was no "winner" in this conflict. Had the UN followed its original objective of driving out the Communist aggressors and re-establishing the 38th parallel, it might be argued that the UN Forces were the winners. Even so, there is no evidence of the Communists being so decisively defeated that they had to look for an end to the Korean War.

The Second Order Change Theory cannot be used as there is no evidence to show the leaders were made to "undergo a psychological process," forcing them to see the problem from a higher or second order paradigm, i.e., threat to their political career or the existence of the state itself. The Cost/Benefit theory is able to explain why the Communists and the United States were eager to end the conflict. However, this theory cannot explain why the armistice negotiations dragged on for more than two years. Even with the ongoing negotiations, battles were fought along the 38th parallel to secure more territories. In fact, just before the signing of the armistice, the Communists launched a final offensive and managed to capture a few outposts from the UN Forces.

The Hawk/Dove Theory does not explain why the Communists and the United States were eager to end the conflict but it does explain why the negotiations were dragged on for more than two years before the armistice was finally signed on July 27, 1953. Stalin was the "Hawk" who refused to allow the Chinese to compromise during the negotiations and had hidden interests in prolonging the conflict. Although none of the theories can, on its own, adequately explain the termination of the Korean War, a combination of the Cost/Benefit and the Hawk/Dove theories however, can explain it.[6]

In the end, the Korean conflict was fought to stop the expansion of communism. A Pollyannaish concept? Perhaps. Regardless, some 60 years later the conflict has never terminated, and the resolution of how to proceed as a society along the demilitarized zone has never come to fruition. For those who have considered the issue of conflict termination at the strategic level, the Korean War has often provided a common basis for discussion of problems inherent in the process. At the operational level, the Korean case also brings to light many of these requirements for war termination doctrine.

There was really no plan for terminating the war in Korea before it started. The Korean War had clearly entered its terminal phase by June 1951; by that date, an informed, objective observer could certainly have predicted the general outline of the eventual outcome.[7] MacArthur's brilliant stroke at Inchon in September 1950 had given UN Forces the upper hand and had prompted an upward revision in U.S. war aims from restoration of the status quo ante bellum along the 38th parallel toward reunification of the entire peninsula under South Korean control. Pursuing this expanded objective triggered massive Chinese intervention in November, prompting MacArthur's laconic comment, "We face an entirely new war."

In the parlance discussed thus far, a cease-fire stopped the fighting on July 27, 1953. There was an armistice signed by North Korea, China, and the UN but not South Korea. Korea is still split into North Korea, which is communist, and South Korea which is noncommunist. The border, protected by a demilitarized zone, was established along the 38th parallel.

Before the armistice, talks had gone on for nearly two years. Eisenhower had promised that if he was elected in the election of 1952, he would go to Korea and end the war. There was no simple way to end the conflict. Talks had collapsed in October 1952. In 1953, the United States threatened to bomb China, but eventually a cease-fire was declared between UN Forces and Korean/Chinese Forces. The Demilitarized Zone, which designates the border between North and South Korea, has remained one of the most heavily-armed stretches of land on Earth; this, in essence, is the conflict resolution situation for the Korean war—an obvious unfavorable ending. Today the stability of the region is threatened by the development of nuclear weapons by North Korea and a constant point of friction along the border. The rest is history.

SUEZ CRISIS . . . KNOW YOUR NEIGHBORS

The 1956 Suez Canal

> . . . operation was launched without a good idea about termination and what the postconflict situation would look like. What if landing at the Suez

Canal at Port Said and Port Fuad did not force the Egyptian President Gamal Abdel Nasser to step down? Were France and Britain then willing to march on Cairo? Would they have international support for such a move? If they seized Cairo, what would the new Egyptian government look like? Could it stay in power without keeping British and French troops in Egypt for years to come? Would the British and French have world opinion on their side for such an occupation?[8]

In the end, the operation did not turn out as planned. The choice of how to terminate the war and bring on a conflict resolution was flawed in the planning process leading up to the war. The United States and Soviets, along with world opinion, forced the British and French to withdraw. President Nasser, rather than being defeated, became the victor and the leader of the Arab cause, while the British and the French lost prestige and influence. How could rational decision makers get it so wrong?[9]

In 1955, Egyptian President Nasser began to import arms from the Soviet Bloc to build his arsenal for the confrontation with Israel. In the short term, however, he employed a new tactic to prosecute Egypt's war with Israel. He announced it on August 31, 1955: "Egypt has decided to dispatch her heroes, the disciples of Pharaoh and the sons of Islam and they will cleanse the land of Palestine. . . . There will be no peace on Israel's border because we demand vengeance, and vengeance is Israel's death."

These "heroes" were Arab terrorists, or fedayeen, trained and equipped by Egyptian Intelligence to engage in hostile action on the border and infiltrate Israel to commit acts of sabotage and murder. The fedayeen operated mainly from bases in Jordan, so that Jordan would bear the brunt of Israel's retaliation, which inevitably followed. The terrorist attacks violated the armistice agreement provision that prohibited the initiation of hostilities by paramilitary forces; nevertheless, it was Israel that was condemned by the UN Security Council for its counterattacks.

The escalation continued with the Egyptian blockade of the Straits of Tiran, and Nasser's nationalization of the Suez Canal in July 1956. On October 14, Nasser made clear his intent: "I am not solely fighting against Israel itself. My task is to deliver the Arab world from destruction through Israel's intrigue, which has its roots abroad. Our hatred is very strong. There is no sense in talking about peace with Israel. There is not even the smallest place for negotiations."

Less than two weeks later, on October 25, Egypt signed a tripartite agreement with Syria and Jordan placing Nasser in command of all three armies. The continued blockade of the Suez Canal and Gulf of Aqaba to Israeli shipping, combined with the increased fedayeen attacks and the bellicosity of recent Arab statements, prompted Israel, with the backing of Britain and France, to attack Egypt on October 29, 1956.

Israeli Ambassador to the UN Abba Eban explained the provocations to the Security Council on October 30:

> During the six years during which this belligerency has operated in violation of the Armistice Agreement, there have occurred 1,843 cases of armed robbery and theft; 1,339 cases of armed clashes with Egyptian armed forces; 435 cases of incursion from Egyptian controlled territory; 172 cases of sabotage perpetrated by Egyptian military units and fedayeen in Israel. As a result of these actions of Egyptian hostility within Israel, 364 Israelis were wounded and 101 killed. In 1956 alone, as a result of this aspect of Egyptian aggression, 28 Israelis were killed and 127 wounded.[10]

One reason these raids were so intolerable for Israel was that the country had chosen to create a relatively small standing army and to rely primarily on reserves in the event of war. This meant that Israel had a small force to fight in an emergency, that threats provoking the mobilization of reserves could virtually paralyze the country, and that an enemy's initial thrust would have to be withstood long enough to complete the mobilization.

Earlier, President Dwight Eisenhower had successfully persuaded the British and French not to attack Egypt after Nasser nationalized the Suez Canal. When the agreement on the canal's use proved reliable over the succeeding weeks, it became more and more difficult to justify military action. Still, the French and British desperately wanted to put Nasser in his place and recapture their strategic asset.

The French concluded, however, that they could use Israel's fear of Egyptian aggression and the continuing blockade as a pretext for their own strike against Nasser. The British couldn't pass up the chance to join in. The three nations subsequently agreed on a plan whereby Israel would land paratroopers near the canal and send its armor across the Sinai Desert. The British and French would then call for both sides to withdraw from the Canal Zone, fully expecting the Egyptians to refuse. At that point, British and French troops would be deployed to "protect" the canal.

When the decision was made to go to war in 1956, more than 100,000 soldiers were mobilized in less than 72 hours and the air force was fully operational within 43 hours. Paratroopers landed in the Sinai and Israeli forces quickly advanced unopposed toward the Suez Canal before halting in compliance with the demands of England and France. As expected, the Egyptians ignored the Anglo-French ultimatum to withdraw since they, the "victims," were being asked to retreat from the Sinai to the west bank of the Canal while the Israelis were permitted to stay just 10 miles east of the Canal.

On October 30, the United States sponsored a Security Council resolution calling for an immediate Israeli withdrawal, but England and France vetoed it. The following day, the two allies launched air operations, bombing Egyptian airfields near Suez.

Given the pretext to continue fighting, the Israeli forces routed the Egyptians. The Israeli Defense Forces (IDF) armored corps swept across the desert, capturing virtually the entire Sinai by November 5. Former U.S. Ambassador Parker Hart said, "We had intelligence reports that many of the Egyptian troops just took off their shoes and ran barefoot to get out of there faster."[11] That day, British and French paratroops landed near Port Said and amphibious ships dropped commandoes on shore. British troops captured Port Said and advanced to within 25 miles of Suez City before the British government abruptly agreed to a cease-fire.

The British about-face was prompted by Soviet threats to use "every kind of modern destructive weapon" to stop the violence and the U.S. decision to make a much-needed $1 billion loan from the International Monetary Fund contingent on a cease-fire. The French tried to convince Britain to fight long enough to finish the job of capturing the Canal, but succeeded only in delaying their acceptance of the cease-fire.

Though their allies had failed to accomplish their goals, the Israelis were satisfied at having reached theirs in an operation that took only 100 hours. By the end of the fighting, Israel held the Gaza Strip and had advanced as far as Sharm al-Sheikh along the Red Sea. A total of 231 Israeli soldiers died in the fighting.

President Eisenhower was upset by the fact that Israel, France, and Great Britain had secretly planned the campaign to evict Egypt from the Suez Canal. Israel's failure to inform the United States of its intentions, combined with ignoring American entreaties not to go to war, sparked tensions between the countries. The United States subsequently joined the Soviet Union (ironically, just after the Soviets invaded Hungary) in a campaign to force Israel to withdraw. This included a threat to discontinue all U.S. assistance; UN sanctions, and expulsion from the UN. U.S. pressure resulted in an Israeli withdrawal from the areas it conquered without obtaining any concessions from the Egyptians. This sowed the seeds of the 1967 war.

One reason Israel did give in to Eisenhower was the assurance he gave to Prime Minister David Ben-Gurion. Before evacuating Sharm al-Sheikh, the strategic point guarding the Straits of Tiran, Israel elicited a promise that the United States would maintain the freedom of navigation in the waterway. In addition, Washington sponsored a UN resolution creating the United Nations Emergency Force (UNEF) to supervise the territories vacated by the Israeli Forces.

The war temporarily ended the activities of the fedayeen; however, they were renewed a few years later by a loosely knit group of terrorist organizations that became known as the Palestine Liberation Organization (PLO). A successful conflict resolution would have caused the destruction of, or at least the marginalization of the PLO as a terrorist organization. However, there was clearly a failure to achieve conflict resolution for the PLO has been a terrorist organization since the war in 1956.

Military leaders are not the only ones who are sometimes guilty of designing wars as if they had to build a bridge that spans only half a river; civilian leaders, too, may order the initiation of a military campaign without being troubled by the fact that they have no plan for bringing their war to a close.

In 1956, Prime Minister Anthony Eden convinced himself that Britain's foreign interests would be acutely threatened if Colonel Nasser retained his control of the Egyptian government. Eden saw in Nasser a second Hitler and decided that he (Nasser) needed to be removed from power. In a highly secretive meeting with the French, he planned a British-French attack on Egypt, coordinated with the Israeli attack on the Sinai.

Eden and the French leaders gave a great deal of thought to the initial military operations, considering where the British and French forces should land, how to time the air strikes, and order their military planners to ensure that casualties stayed very low. They also carefully designed various public statements and diplomatic maneuvers in the United Nations, particularly to avoid giving the impression of collusion with Israel. To comply with these political constraints for the initiation of the war, the military planners were ordered to choose Port Said as the landing site, instead of Alexandria which they favored. Alexandria would have offered better access to Cairo, a highly important consideration for ending the war.

Prime Minister Eden, with all the careful attention he bestowed on reducing the costs and risk of the war's beginning, neglected to plan for an ending that would have accomplished his war aims. How could the landing in the Suez Canal area bring about a situation in Egypt that would have resulted in the overthrow of Nasser? If British and French Forces would have had to march on Cairo to depose Nasser, how much time would this have required and what would be the losses and risks to stay in power short of a prolonged stationing of British and French troops in Egypt?

In the end, how far concessions were meant to go can be seen in British policy on the issue of the Suez Canal. In the early 1950s, Britain still kept troops stationed along the Suez Canal in accordance with a treaty with Egypt dating back to 1936. The Egyptians exerted mounting pressure against the British Forces on their soil, including sporadic guerrilla warfare. But in 1954 the British government agreed to evacuate its troops and through this concession—which some British parliamentarians thought shameful "appeasement"—vastly improved its relations with Egypt. Premier Nasser called it "a good agreement on which we can start at once to build a new basis of relationship with Britain and the West." A possibly prolonged war between British troops and Egyptian guerillas had thus been prevented.

However, the improvements in Anglo-Egyptian relations came to naught. The British still had a major interest in the Suez Canal Company, whose lease from Egypt for the exclusive right to operate the canal had another 14 years to run.

VIETNAM . . . WIN THE BATTLES; LOSE THE WAR

In general, without clear objectives it is impossible to develop a clear understanding of what the endstate should look like and, therefore, it becomes difficult to develop a functioning exit strategy for conflict termination. A failure to clearly define political and military objectives is probably one of the most severe criticisms one can make of the Vietnam War. Recent operations indicate that this lesson may have only been partially learned.

The Vietnam War is considered to be a low point in the history of the U.S. military. The U.S. military was pulled out from the theater of war hastily in line with the revised U.S. policy of disengagement, leaving South Vietnam to finally fall to communist North Vietnam in 1975. The U.S. military had deployed more than half a million people at the height of the war. One strategist claims that the failure of the American war effort was because of limited commitment by the political leadership due to the strong anti-war movement.[12] In contrast, the military was totally committed to the objectives, even though it had already suffered casualties in pursuit of the objectives. The question becomes, was the other tier of the Clausewitzian trinity the policy makers and the public?

The U.S. military was unable to win the war outright as it was never given a free hand to wage the war due to the political costs of the war. In essence, the political leadership never defined the war termination objectives or conflict resolution requirements to achieve an endstate derived in its introduction of forces in 1963 with a direct aim of stopping the South falling into "communist" hands.

With little criteria to measure success once forces were introduced, it fell on the political leadership to plan for a minimally satisfactory war ending, using the limited military assets and abiding by a time frame that was acceptable to the public. The decision to end the war lay in the political leadership, but President Lyndon Johnson was unwilling to admit that the war could not be won. Thus, the war dragged on inconclusively for several years without any victory in sight. This is the age-old problem with not having a clear set of defined war termination and conflict resolution plans—the military and, in specificity, the leadership of the military is left holding the bag to define success.

The Vietnam War, ending in an armistice agreement, had a very long and problematic war termination process. Although the introduction of forces into Southwest Asia was done with little thought to an achievable endstate and little direction in an attempt to define victory, allied influence, unsynchronized U.S. military and political objectives, and pressure from public opinion contributed to an unsuccessful war termination conclusion. At the international level, the Soviet Union and China provided significant military aid to North Vietnam, helping the North Vietnamese achieve a military advantage against the United States leading into the 1972 Tet offense.[13]

These alliances, however, also advised North Vietnam to pursue a diplomatically based war termination agreement rather than military escalation; however, Hanoi would not accept a cease-fire without an agreeable political solution and proceeded with the Tet offensive in 1972, with the Politburo believing that the offensive would alter both the military and political balance of the war at this point. Although Hanoi did not achieve a decisive change in military balance, the offensive prompted diplomatic negotiations toward an armistice agreement.[14]

At the domestic level, the civil-military strife between President Lyndon B. Johnson and Robert McNamara with their Joint Chiefs of Staff was well known in the early part of the Vietnam War. The Joint Chiefs believed in the use of overwhelming military force while McNamara adopted the "graduated pressure" of limited and slowly escalating warfare.[15] At this point, the policy makers and the military leaders for the United States were at odds over what to do about Vietnam.

To add to the difficulties of trying to define victory well after the introduction of combat power to the equation in Vietnam, public opinion through anti-war demonstrations pressured U.S. policy makers to end the war as soon as possible, and this impacted overall national morale and support toward the war, and ultimately, the U.S. will to fight. However, in the case of North Vietnam, the lack of public influence in this autocratic (or totalitarian) regime actually served to prolong the negotiation process and war termination. Vietnam War termination events suggest that "policy makers on both sides of a conflict made their decisions for war or peace on the basis of the military balance and the costs and constraints of battle."[16] However, in the absence of clearly defined political goals on the side of the United States, the military objectives probably had no good chance of success, and civil-military strife made that gap even wider.[17]

So although there was clearly an objective ambiguity with ending the war in Vietnam, there were some key successes that shed some light on civil participation in the conflict resolution effort. A number of organizations provided early nonmilitary assistance to the Republic of Vietnam, including the International Cooperation Administration and the Development Loan Fund—the same entities created to implement the Marshall Plan in Europe after World War II. As early as June 1955, resources from these organizations were directed toward land reform programs and training for South Vietnamese police forces and intelligence services in counterinsurgency tactics.

In 1961, President John F. Kennedy created the U.S. Agency for International Development (USAID), an independent federal agency that received its policy guidance from the State Department; clearly a lasting legacy as well as a lasting problem. Between 1962 and 1975, thousands of USAID workers were embedded throughout South Vietnam, helping to establish schools and medical treatment facilities, build transportation and utilities infrastructure, and administer many other forms of nonmilitary assistance to the Vietnamese people.

Unfortunately, management of these largely civilian-run programs was not well coordinated in Saigon with senior embassy and military command officials. As part of the pacification/counterinsurgency plan, Kennedy also increased the number of U.S. military advisers working with the Army of the Republic of Vietnam and local defense forces.

When President Lyndon B. Johnson met with the heads of the South Vietnamese governments in Honolulu in February 1966, pacification was stated as one of three components of the Johnson administration's strategy, the other two being military pressure and negotiations. On his return to Washington, Johnson appointed Robert W. Komer as a special assistant for pacification to Vietnam. Komer worked tirelessly in Washington for the remainder of 1966 to subordinate all pacification organizations and activities to General William C. Westmoreland's Military Assistance Command Vietnam (MACV). In the face of determined opposition from Secretary of State Dean Rusk and Ambassador Henry Cabot Lodge, Johnson set a 90-day deadline for a civilian solution to the organizational problem.

The solution, the Office of Civil Operations, was created in November 1966. It combined the personnel and activities of USAID and several other civilian organizations. It employed about 1,000 American civilians and directed a program budget of $128 million and 4 billion South Vietnamese piastres. Meanwhile, Westmoreland created within his MACV headquarters a Revolutionary Development Support Directorate and named a general officer as its director; again a military oversight over a clearly civilian skill set. The Office of Civil Operations was short-lived—it was succeeded in early May 1967 by the Civil Operations and Rural Development Support (CORDS) program. The first director of CORDS was Komer, who enjoyed direct access to Westmorland, through his status as a MACV deputy, and to U.S. Ambassador Ellsworth Bunker, through his appointment at ambassadorial rank.

The CORDS program was implemented through a command and control structure that paralleled or was intertwined with the military command structure down to the province senior adviser level. In addition to all pacification activities, CORDS was also responsible for providing advice and support to the South Vietnamese militia, conducting the war against the enemy's clandestine politico-military command administrative infrastructure (the PHOENIX program), and coordinating with the South Vietnamese government for recovery after the 1968 Tet Offensive.

When Komer accepted the offer from Johnson to become U.S. Ambassador to Turkey in October 1968, he was succeeded as director of CORDS by William Colby. Colby and the new MACV commander, General Creighton Abrams, both strongly believed in the CORDS mission and worked cooperatively toward its accomplishment until the final withdrawal of U.S. troops from Vietnam in early 1973. The feeling of many who participated in the program was that it had been highly effective, but came too late to alter the war's outcome. Because as defined in the early part of this section, no matter how good a plan is to terminate a war or provide a feasible conflict resolution plan, if the endstate or objective is flawed no plan can be expected to succeed.[18]

Therefore, although there were some early successes in the implementation of a number of civilian oversight programs to help bring the conflict in resolution to a successful end, to include USAID which continues to support postconflict efforts today, there was clearly a void in establishing a universal plan to achieve success in terminating the war and establishing a peaceful Vietnam in the wake of the 13-year fight. Vietnam and its historically known lack of a plan for success may be the most notorious example of a failure in achieving an endstate in conflict.

SOMALIA . . . FEED THE MASSES AND THEN LEAVE WHEN IT GETS TOUGH

Many ask the question: What happens when a nation decides to pull out of a conflict short of accomplishing its goals? The prevalent military thought in answering this question seems to be that trying to disengage or even refuse to participate can cost a nation like the United States credibility. The consensus seems to be that once this happens, other countries will question the U.S. commitment and deny their future support in discords that matter to the United States. However, there are several historical examples indicating that few if any nations suffered loss of credibility when they decided to act in their best interest. Although the U.S. reluctance to extricate itself from Vietnam hurt U.S. credibility worldwide, on the contrary, France increased its credibility by leaving the war in Algeria in the early 1960s. Somalia in the early 1990s is a totally different case.

Few could argue that Somalia is a failed state and many blame its failure as a nation, in part, due to the failed exit strategy executed by the United States in 1993. The objectives for the employment of combat forces, even under the auspices of the UN, were ill-defined from the start. Additionally, *mission creep* reared its ugly head as events unfolded in the troubled nation. Today we call mission creep *consequence management*. My premise is that regardless of what it is called, it exists because of poor planning; once a plan is poor, unforeseen consequences occur, leaving a military force and its supporting agencies struggling for a solution. Therefore, it is clear that if we don't plan for war termination and conflict resolution in some detail from the onset of hostilities, the chances of it resulting in failure exponentially increase over time.

In an attempt to appease the international cry to support a massive humanitarian effort, President George Bush issued the order to commence Operation Provide Relief to airlift food to Somalia in an attempt to arrest the widespread starvation and lessen the obvious suffering. In May 1992, the operation became United Nations Operation in Somalia (UNOSOM II) that included an attempt in extensive nation-building efforts. The initial requirement and request was to support with logistical supplies; however, President Bill Clinton, now in office, agreed to provide a quick reaction force consisting of combat forces to include special operations units. The expanded mission turned into a misguided effort to hunt down warlords resulting in the loss of 18 U.S. soldiers, a disenfranchised American public, and ultimately in the withdrawal under fire of all U.S. Forces some five months later.

In the end, the intended involvement in Somalia was expected to last just a few months and to establish a secure environment for private relief groups to deliver assistance to starving Somalis. The idea of saving people from starvation appealed to the United States and made Americans feel less helpless about the atrocities occurring in the former Yugoslavia. Arguably, the actual mission was never really seen as definable or even doable where the United States could rapidly make a significant difference. The State Department never hid the fact that maintaining a secure environment might prove difficult. In reality, the clan violence was just too much for the small combat force and dysfunctional UN force supporting the humanitarian relief to handle.

The final result left Somalis at the mercy of the warlords and the nation a failed state. That, in essence, was the conflict resolution. Although many think that the conditions required to achieve an endstate for what was once termed operations other than war (but translates better into all operations besides high-intensity conflict) are difficult to define and require continued refinement during the operation, there is no excuse for ill-defined goals and poor planning to establish criteria to achieve even the poorest of defined objectives. Operation Restore Hope was a typical foray as an operation "to just do something," especially in the light of the U.S. inability to affect the situation in Bosnia.[19]

Ultimately, Somalia was not a vital interest for the United States and, in spite of the good feelings that operation initially generated, Americans viewed their involvement as limited in scope and without a clear path established early on to achieve resolution to a war whose objectives seemed to continually change. Adding in the coalition and UN who each had their own understanding of what a restored Somalia should look like and what they were willing to contribute make the effort impossible to define mutual endstates, war termination, and conflict resolution—if not defined and agreed upon at the beginning of the conflict.

BOSNIA FOR A YEAR. . . . OR MORE

History has shown that predetermined timelines habitually do not work. One of the major problems in Bosnia was that the military habitually finds it difficult to plan how to employ armed forces effectively and then have a viable strategy to leave given a strict timeline. The lack of clearly defined objectives or endstates makes it tough for the military to know where to begin and where to end. Given this ambiguity, how could we have possibly gotten a postconflict strategy in the Balkans?

On December 18, 1997, President Bill Clinton announced that he had decided in principle to keep U.S. military forces in Bosnia past the June 1998 deadline and into the indefinite future. Although the decision to maintain a force in the fight was the right call, any strategist would tell you the employment and maintenance of a combat element is dependent on events and not time. The fact of the matter is that, regardless of the goals established early on, to tag a timeline to its accomplishment can only work if the endstates, war termination strategies, and conflict resolutions are clearly defined and, perhaps most importantly, resourced. They were not in Bosnia.

Clinton finally admitted that to achieve success in Bosnia, the forces employed would have to achieve a series of "benchmarks" toward creation of a "self-sustaining, secure environment" in Bosnia and that this goal would most likely not be reached by the time he left office in 2000. In the meantime, the administration and military leadership continued to struggle to define what benchmarks would eventually be used to determine progress. Among several sample standards Clinton offered at the time were whether Bosnia's governing institutions—in which power is shared among Bosnian Serbs, Croats, and Muslims—would be strong enough to survive on their own after a NATO pullout, whether the country's civilian police force was capable of maintaining order on its own, and whether the military was clearly under democratic civilian control.[20]

It was a more than a month into the air campaign against Serbia over Kosovo before the international community addressed the issues of termination objectives. General Wesley Clark cited that the endstate and objectives can slip and change, especially if the endstate is not clear. That meant that if war starts, like the one in Bosnia, with no set termination and resolution criteria in place, it is more apt to turn for the worse. Without planning ahead of time for what comes after the fight, the potential for escalation, mission creep, consequence management, and hasty decisions leading to extension or premature withdrawals becomes greater.[21]

Although timelines themselves lead to inappropriate managing of resources, namely soldiers, sailors, marines, and airmen, there is some credence to at least stating that an effort to achieve an endstate will take some time. The one-year-and-out promise by Clinton during the initial foray into the Balkans still remains a point of contention when reviewing the history of war termination strategies.

GLOBAL WAR ON AN IDEOLOGY

If the changing face of battle is true; meaning that the enemy is elusive and conflict itself is based on the rise of globalization leading to the ascendency of transnational issues like terrorism, nuclear proliferation, poverty, and migration, it would seem right that our effort to wage a war against a worldwide "network," such as terrorism, would make sense.[22] Many believe that when a nation chooses to participate in low-intensity actions, it should establish or be given well-defined goals and objectives. However, even specific goals may not necessarily translate into preplanned exit strategies or a clearly identifiable conflict termination. Sometimes the goal may be never to end the fight; many would argue that was George Bush's intention in the 2001 administration's announcement of its ideologically driven Global War on Terror (GWOT).

The wars in Iraq and Afghanistan are a quest to rid the world of an attitude rather than stomping out a well-defined enemy, which leads to the ambiguities in both OIF and OEF. Many would say that GWOT has no war termination strategy; arguably, the goal is to continue to fight for the sake of fighting.

Terminating a deliberate act of violence in the form of combat against a foe has been increasingly more difficult to conclude. Earlier in this text, I supported GWOT as a "good" example because it is obvious that the war efforts in both Iraq and Afghanistan have wreaked havoc on the various terrorist networks associated with those theaters; however, it is unlikely that success in both OIF and OEF will stamp out the terrorist effort in other parts of the world. Even if the Taliban and Al Qaeda are destroyed in detail (which is unlikely even in Iraq and Afghanistan), their destruction does nothing but perhaps minimally deter the likes of other terrorist organizations that exist throughout the world including Colombia's National Liberation Army, Revolutionary Armed Forces of Colombia, and United Self-Defense Forces/Group of Colombia (Autodefensas Unidas de Colombia); Georgia's Zvuadist; the Palestanian Liberation Front; Japan's Aum Supreme Truth; Hamas (Islamic Resistance Movement); and Sri Lanka's Liberation Tigers of Tamil Eelam just to name a few.

There is no doubt that it is easy to cause havoc and destruction; but the destruction of one terrorist element, no matter how significant it is to the world of terrorism globally, at the end of the day is only one terrorist element in a never-ending pool of radicals in the world that have an issue with authorities or are fanatically incessant about some religious or ideological principle.

AFGHANISTAN

The combat phase of OEF was an unqualified success. By all accounts, the opening stages of the fight brought the Taliban to its knees. My experience with the 75th Rangers and the Joint Special Operations Command at the time gave me first-hand situational awareness of the wrath brought on the enemy in the U.S. quest to destroy their ability to both fight and rule the nation of Afghanistan. But when the high-intensity conflict transitioned into conflict resolution, activities required a definitive counterinsurgency (COIN) and foreign internal defense operation to set the path for success in Afghanistan. However, the fact is that defeating terrorists in a counterinsurgency does not work in a myopic way—it didn't in Iraq for the first four years and it won't in Afghanistan.

As it relates to war termination and conflict resolution strategy, how can we expect it to be a feasible course of action? What does it say to the people of a nation when the "occupiers" are there just to defeat an enemy, and in this case, an enemy who is imported? And when they are done, they just leave the nation in a situation where the political environment is vulnerable to the upsurge in violence.

In 2009, Presidential Candidate Barack Obama correctly argued that when the United States prematurely turned away from Afghanistan to focus on Iraq in 2002 to 2003, the result was the near-collapse of the new Afghan government and the resumption of widespread civil strife. The administration and the U.S. military is now reeling from a lack of focus early on—no plan was devised for accomplishing an endstate in OEF and to add to the confusion as to how to achieve war termination and conflict resolution in one theater, another one, Iraq, begins its wayward path toward success.

Even Ken Pollack, author and Middle East expert, agrees that to focus on Afghanistan alone and to turn away from Iraq prematurely would have dire consequences for Iraq, whose fragile government will be more likely to fail, and for the United States, because success in Iraq is vital to U.S. interests. At the time of this writing, there was no doubt that the situation in Afghanistan was getting worse. Some eight years after the war began in Southern Afghanistan, civilian casualties had increased 40 percent from 2007 to 2008. According to one report, the death toll—2,118 civilians killed in 2008, compared with 1,523 in 2007—is the highest since the Taliban government was ousted in November 2001, at the outset of a war with no quick end in sight. Civilian deaths have become a political flashpoint.[23]

Surely we have learned in Iraq that the population needs to be protected; even, in some cases, from themselves.

Civilian deaths in Afghanistan have eroded public support for the war and inflamed tensions with President Hamid Karzai, who, at times, has bitterly condemned the U.S.-led coalition for the rising toll. President Obama's decision to deploy more troops to Afghanistan, although the correct decision, raises the prospect of even more casualties.[24]

Regardless, in December 2009, Obama announced that the United States would deploy an additional 30,000 to 40,000 troops to Afghanistan, but would seek to remove them in roughly 18 months, with the expectation that the Afghan National Army would then be ready to take on the task. The implicit expectation is that the United States will be able to help Afghanistan as it helped Iraq. Even though Afghanistan and Iraq are very different countries facing very different problems, if the United States achieves its goals in Afghanistan, it may nonetheless face some of the same problems of premature withdrawal. As senior U.S. officers in Iraq regularly intone, progress in these kinds of wars "doesn't mean no problems, it means new problems."[25]

But Afghanistan is not Iraq. We are so loathe to be accused of fighting the last battle, that those words have almost become a credo in OEF. An officer who dares say, "When I was in Iraq" risks the professional scorn of his peers. Policy makers have made the use of the word *surge* so taboo that *civilian uplift* is favored despite the snickers that inevitably follow use of the phrase. Pundits in Washington, naturally eager to compare simultaneous operations, almost universally believe that to point out too many similarities puts them in an intellectually inferior position.

Although Afghanistan is NOT Iraq, it should go without saying that the much-touted *Counterinsurgency Manual*, developed and used extensively in Iraq, applies also in Afghanistan. If it didn't, it would not stand as an essential part of the body of military doctrine. It therefore follows that application of that doctrine, or failure to implement that doctrine effectively, has carved grooves in the personal and professional makeup of those officers who served in Iraq and who are now serving in Afghanistan.

> U.S. forces committed to a COIN effort are there to assist a HN [host-nation] government. The long-term goal is to leave a government able to stand by itself. Achieving this requires development of viable local leaders and institutions. U.S. forces and agencies can help, but host-nation elements must accept responsibilities and grow in new areas to win the support of their own people. While it may be easier in the short term for U.S. military units to conduct operations themselves, it is better to work to strengthen local forces and institutions and then assist them.[26]

This is not to say that the specific strategy in Afghanistan should be built on the strategy of Iraq. Nor is it to say that implementation of first principles of counterinsurgency should be the same in both places. It *is* to say that if you dig down into a specific area of OEF—as an example, host-nation capacity building—you will find similar challenges and therefore corrections in an organization that is adaptable as the Army has proven the case in this near-decade of counterinsurgency. The quicker we recognize how to assist and partner, leading from "behind, beside and beneath,"[27] and *believing* that those indigenous force can do *it*,[28] the more quickly we will successfully transition, leaving behind a durably stable country able to eventually partner with the United States in pursuit of mutual, regional, and global interest.

Although it is instructive to examine where we are in the fight in Afghanistan, arguing that the war is at a turning point with the intent to demonstrate lessons captured from Iraq and from the earlier years in Afghanistan, what to do now is not the point of this work. What should have been done, or more importantly, what wasn't done in the early stages of the fight in OEF is the focus of this study.

One area in particular that most focus on when defining the initial stages of a prewar planning process is that conflict resolution is all about capacity building at the local level along three lines of operation: security, governance, and development. Getting this right in both the planning and execution phase is fundamental to successful transition to host-nation control. The problem we have in Afghanistan is that this process or focus didn't occur in detail to include the commitment of resources until 2009 when General Stan McChrystal was given the charter to achieve success; some 8 years after the war had started. The 2010 replacement of McChrystal with General David Petraeus, while potentially a great risk right when there appeared to be a notable transition in both policy and execution of a feasible COIN operation, has proven thus far to have nothing but a positive impact.

Afghanistan is an example of a sudden exigency for the resources and skill sets to conduct nation-building. The requirement to understand how to plan and execute nation-building is amplified because of these ongoing conflicts, the incessant fear of terrorism that could lead to conflict, and a need to assist in postconflict resolution.

In this instance, the United States has become deeply involved, but not merely for the sake of Afghanistan. Instability in countries such as Afghanistan threatens not only a nation's own existence, but also the regions around them and, at times, other parts of the world. "The United States has rightly set itself the mission of ensuring that Afghanistan is not a safe haven in which Islamic extremists can locate academies for anti-American terrorists," said Ashton Carter in an interview. "It is another matter altogether to take on the mission of turning Afghanistan into a civil society."[29] The question is whether the government of Afghanistan and the people of Afghanistan are winning, because ultimately, it's their victory that matters.[30]

My argument is that if we had established this direction early on, i.e., in September 2001, the past eight years of fighting and the loss of blood and treasure would have been spared or at least diminished.

Prior to the Obama administration's decision to surge in Afghanistan, strategist expert Dr. Schake from the Hoover Institute wrote that "more American troops aren't enough to succeed in Afghanistan. What else needs doing depends on why you think the Taliban have gained ground in the past 18 months [mid-2007 to early 2009]? Is it because we have too few troops to hold areas that have been cleared of Taliban influence? Is it because Afghans are fundamentally sympathetic to Taliban aims? Or are Afghans so downtrodden from the terror and distrustful of American staying power they won't stand up and help?"[31]

Within the global conflict against terror, each counterterrorist campaign such as OEF is simultaneously a limited conflict and an unlimited conflict. The limited conflict is with the regime in power. The military objective then becomes regime change. The unlimited conflict is with the terrorist organizations. In a conventional military confrontation with either entity, victory is a foregone conclusion. U.S. training and technology alone can bring the enemy to its knees in the early hours of a fight or during major combat operations.

The decapitation of a nation-state's leadership—the Taliban in Afghanistan or Saddam in Iraq, creates a civil government vacuum that allows terrorists and bandits to operate at will. Moreover, a country's descent into anarchy could generate the despair that will entice more individuals to join criminal and terrorist organizations. Therefore, in recognition of this potential, the operational design of a campaign, especially when it involves a GWOT aspect, should focus on a military endstate in which the conditions have been created that allow security operations to transfer to indigenous forces. This is key to planning in advance.

When one reads headlines like, "The Pentagon's senior military leaders are worried that the security situation in Afghanistan is stalemated or deteriorating, and now are preparing a far-reaching plan that would prepare the U.S. military for a war that could last three to five more years, officials said," it begs the question, "What is the overall plan or endstate, or are we figuring it out as we go along"?[32] Without this focus, planners lose sight of the decisive leverage and then play a game of catch up. This is precisely what occurred in Iraq and Afghanistan.

Other measurable criteria include measuring the people's support for their government and security forces, as well as support for one level to the next, through many indicators. These indicators are participation of the people in selecting their representative councils and leaders; turning to their government to resolve issues and grievances; resorting to the police and legitimate local security apparatus for protection from violence, intimidation, illegal taxation, and insurgent justice; and taking ownership of community projects, service delivery, and other expenditures of resources. The need for a capable Afghanistan Army and police force is critical, but to define it as such, meaning truly engaging both U.S. and Afghan leadership with the need to resource and commit to providing a capable indigenous force some nine years after the bullets started firing is suboptimal in the very least.

FUTURE FIGHTS

An ancillary "bad" effect on some of these misguided and unbounded conflicts is their potential for spillover to other fights; either because of poorly defined exit strategies, as seen initially in Iraq and now in Afghanistan, or because of our inability to see that *mission creep* is an often unwelcomed byproduct of an inability to see the outcome of a fight. The conflict in Afghanistan, even more so than in Iraq, has been influenced by its border countries or the greater Afghan region. The recognition of this greater regional involvement requires a strategy for engaging places like Pakistan and Uzbekistan to achieve success and an accompanying war termination strategy for Afghanistan. Establishing a process or at least understanding it for this war and its entire area of interest should help strategists to understand the challenges of border nations and how they affect conflict resolution in the future.

Diplomatic efforts with countries that possess few vital interests make it more difficult to develop an overall strategy for the region; the fight in Afghanistan is a very good example of this for there are few natural resources in that nation to capitalize on when defining the good of coming to resolution in this resource-starved nation. An examination of past efforts such as the Weinberger Doctrine in its dogmatic depiction of what is a vital interest will help strategists to understand the dichotomy involved with having to engage regionally to accomplish a specific goal.

Those that have policy responsibility in the security arena will have to think ahead and work effectively with other international actors. Those who simply react to events will be overtaken by events and those who try to tackle global problems on their own will become overwhelmed.

Of note, understanding how any future U.S. diplomatic or military engagement in Pakistan (or those that should have already occurred) is important to comprehending the conflict resolution strategy in that particular region and is a good example of this ancillary effort for future fights. This is a great opportunity for the United States to define an endstate and establish an exit strategy for whatever it is we decide to do with Pakistan while fighting the war in Afghanistan proper.

Although ancillary effects are based on blurring of boundaries, what is also a keen possibly for the future is the third dimensional facet of war: information and cyber networks. Can the world make a peaceful transition into what we've called a *Third Wave civilization*—a knowledge-based economy and society? If we understand conflicts of the past, they suggest that when a new wave of change brings a new type of economy, new technology, and a new way of life, it conflicts with those whose pocketbook and psyches are hitched to the previous system. When a new wealth system and way of life arise, there are winners and losers.

Today we are moving to a new Third Wave economy and society, so it's hardly surprising that the U.S. military is utterly dependent on computers, information-technology, satellites, advanced communications and "smart weapons." The wars of the future will increasingly be prevented, won, or lost based on information superiority and dominance. And that isn't just a matter of eliminating the enemy's radar. It means waging a kind of full-scale cyber-war. The greatest U.S. weakness is the absence of a global strategy. The United States has the biggest, strongest, most expensive, best-educated, best-equipped military in the world, but it is strategically brainless at the top because many senior advisers and leaders think that agility is a substitute for strategy. Agility is necessary, but it is reactive. If you have no strategy, you are very likely to become part of someone else's strategy a la GWOT.

CHAPTER 5

The Missing Link:
The Interagency Struggle

Over the years, the interagency system has become so lethargic and dysfunctional that it inhibits the ability to apply the vast of the U.S. government on problems. You see this inability to synchronize in our operations in Iraq and Afghanistan, across our foreign policy, and in our response to Katrina.

—Gen. Wayne Downing, Former Commander-in-Chief,
U.S. Special Operations Command

As the argument goes, you can't lead a horse to the water and make it drink if you don't have a horse. Or even worse, what happens if you have a horse, but there is no water? The horse in this analogy represents resourcing; the water (or the water dispensing system) represents the plan to dispense those resources.

We are at a time in history where the "nation," as a whole, will never be at war again. Only small facets of our national government are fighting the fight. As a Tennessee Congressional Representative so eloquently put it a few years ago at a speech at Fort Campbell, Kentucky, "The military is at war and the rest of the nation is at the mall."[1] This chapter is a study of interagency inadequacies in our government. These inadequacies have had a direct impact on the inability to plan for any type of reasonable conflict resolution in time of war.

The United States has a predilection for neat categories of activity and clear divisions of labor. One manifestation of this tendency is emphasis on a clear division between military and political realms and a related belief in a clean separation of military and civilian activities. However, war is a complicated and messy human phenomenon that defies easy categorization. The fundamental political core of war admits to few natural limits. The stakes of war are usually profound, and therefore, the effective remedies can be no less intense.[2]

Although conventional military efforts are necessary and important in counterinsurgency (COIN), they are only effective if integrated into a comprehensive strategy that addresses all relevant societal needs. This requirement is frequently expressed in terms of applying the appropriate instruments of national power to meet those needs. The logical relationship of agency to effort, however, is secondary to the necessary societal outcome. Put another way, solving a problem is more important than who solves it. However, we have seen that there is no doubt that whoever solves the problem often decides if the problem is ever actually solved.

Ideally, a society's needs will be met by those organizations having the most appropriate expertise or comparative advantage in a particular task. Realistically, the counterinsurgents will have to rely on whoever can perform a particular task when and where it is needed rather than standing on formality about who should perform it. Quite frequently (and if are we are totally honest, almost always), of all the COIN representatives it is the armed forces who are present and can act, and most importantly have the resources. Sheer capacity and the logic of one of the most fundamental aspects of warfare—the control of physical space (and the people and material in it)—will often place members of the armed forces at crucial societal nodes. Unfortunately, as we have seen in the most recent fights in Operation Iraqi Freedom (OIF) and Operation Enduring Freedom (OEF), the military oftentimes doesn't organically have the skill sets to solve many of the problems associated with issues such as economic infusion and governance assistance.

Although political, social, and economic programs are most commonly and appropriately associated with civilian organizations and expertise, the salient aspect of such programs is their effective implementation and not who performs the tasks. COIN programs for political, social, and economic well-being are essential elements for supporting local capacity that can command popular support.

In reality, the military can and should be engaged in using its capabilities to meet the local population's fundamental needs, mindful that these needs vary by society and historical context. The military performs a crucial role in creating the security conditions to permit a society to function normally. Principally, security forces should seek to prevent intimidation and coercion by the insurgents while setting the conditions for interagency and other civilian organizations to surge into a capacity-building mode. In the end, in COIN the performance of military and nonmilitary activities is interdependent so we cannot separate what is needed by the military and the civilian sector. The needs of the society under which we work obviously don't make this delineation. Facilitating active support for the host-nation government by the local population deprives an insurgency of its power.

Before the war on terror began, John R. Broule stated that, ". . . [An interagency] element must include members well-versed in the application of military, diplomatic, informational, and economic instruments of national power."[3] In this visionary article, he defined a need to create an interagency organization and practices that can effectively conduct termination planning. To do this, however, requires much planning and an innate understanding of the capabilities that different agencies and organizations may or may not have at their disposal.

There is lots of talk nowadays about *hard power* versus *soft* or *smart power* in the Washington, DC, area. As Secretary Gates said, ". . . these new threats [asymmetric enemy tactics] also require our government to operate as a whole differently—to act with unity, agility, and creativity. And they will require considerably more resources devoted to America's nonmilitary instruments of power."[4] He goes on to say that one of the most important lessons from our experience in Iraq, Afghanistan, and elsewhere has been the decisive role that reconstruction, development, and governance plays in any meaningful, long-term leadership and political will. However, the Department of Defense (DoD) had taken responsibility of the burdens that might have been assumed by civilian agencies of the past, although new resources have permitted the State Department to begin taking a larger role in recent years. Still, forced by circumstances, our brave men and women in uniform have stepped up to the task, with field artillerymen and infantrymen doing things like building schools and mentoring city councils—usually in a language they don't speak. But, as recognized by Secretary of Defense Gates, there is no replacement for the real thing—civilian involvement and expertise; a capability the military just can't bring to the fight.

There is no doubt that interagency coordination throughout military operations is the linchpin. Instead of separating hard from soft powers; it is clear they need to be fully infused and integrated. In *Soft Power: The Means to Success in World Politics*, Joseph S. Nye states, "More than four centuries ago, Niccolo Machiavelli advised princes in Italy that it was more important to be feared than to be loved. But in today's world, it is best to be both. Winning hearts and minds has always been important, but it is even more so in a global information age. Information is power."[5]

MACHIAVELLIAN MANEUVERS FOR THE MODERN MILITARY: HITTING IT HARD WITH THE SOFT POWER

> No enterprise is more likely to succeed than one concealed from the enemy until it is ripe for execution.
>
> —Niccolo Machiavelli, *The Art of War*, 1521

Although clear, concise, and achievable objectives must be defined in establishing a plan to fight a war, terminating it at some point and then reverting to conflict resolution, what is critical in achieving these facets of victory is the ability to first possess and then employ the resources, or in other terms, a means to an end. In essence, to achieve an endstate championed by this book, one first must recognize what it means to employ power in its varied forms.

Niccolo Machiavelli wrote his famous dissertation on power, *The Prince*, in 1517. His thoughts on the rules of power encompass the struggles for every level of power, from the proletariat struggling in the corporate world to strategies performed by the world leader in the 16th century to now. Adolfe A. Berle wrote that *The Prince* is "the greatest single study of power on record."[6]

The philosophies in *The Prince*, known as Machiavellianism, have been viewed as evil throughout the centuries, but as most business leaders and politicians agree, Machiavelli has only defined the physics of power.

> . . . since it is my intention to write something of use . . ., I deem it is best to stick to practical truth of things rather than to fancies. Many men have imagined republics and principalities that never existed at all. Yet the way men live is so far removed from the way they ought to live that anyone who abandons what is for what should be pursues his downfall rather than his preservation, for a man who strives for goodness in all his acts is sure to come to ruin, since there are so many men who are not good.[7]

"Fancies" are what Machiavelli means by the theories that recognized men as good and thus are able to be controlled by good. But in this quote, Machiavelli points out that man does not live in such a fashion. Therefore, those acts that are "other than good" are necessary for acquisition and preservation of power in society. Machiavelli set the precedent for the cold and calculated regardless of the century. He discusses frankly the necessity of cruel actions to keep power. He was in the business of power preservation, not piety. Those who desire power in any situation may look to his strategies for solid aid; ". . . he (the leader of the state) must stick to the good so long as he can, but, being compelled be necessity, he must be ready to take the way of the evil."[8]

The core of Machiavellianism is essentially defined as the priority for the power holder is to keep the security of the state regardless of the morality of the means—to achieve victory at all costs. The answer according to Machiavelli is ". . . that it would be best to be both loved and feared. But since the two rarely come together, anyone compelled to choose will find greater security in being feared than in being good." [9]

With this attitude in tow, true Machiavellians therefore don't require a set of criteria to terminate war and bring it to resolution, because in the end, they, at times, brandish a knife in rogue fashion to defeat an enemy force regardless of the outcome and the collateral damage. In short, Machiavellian leaders tend to do what they need to do to win.

Why this is relevant to this work is because many leaders act in this fashion, ignoring the need for a plan to support a need to surmise a prior conflict resolution plan. But, in reality this type of decision-making process has brought on some of the greatest victories in history, and at the same time, some of the most failed outcomes in the postconflict phase.

For example, true or not, Henry Kissinger has been called the greatest diplomat of our time. He recognized the need for separation of morals from the power struggle; the irrelevance of morality in politics. For his theories he also has been called the Machiavelli of the 20th century. Like a true Machiavellian he, as Secretary of State under the Nixon administration, systematically analyzed the struggle between the democracy of the United States and the threatening communism of China and the Soviet Union. His political ideas were centered in Machiavellian fashion on the issues of ends (what kind of society do we want after the guns stop shooting?) and means (how we will get it?). In that one sentence we can relate Kissinger to Machiavelli because both settled for modern governments the question of ends: conquest, and the question of means: force.

What is our objective? Our means? What is the worst that can happen? What is the best? Henry Kissinger asked himself these questions when he designed the policy of detente with China and the Soviet Union. His objective: to contain the threatening communism.

Looking back further in history, what caused President Harry Truman to drop the atomic bomb on Hiroshima and Nagasaki? The casualties reached approximately 120,000 with the extending effects of radiation. To relate it to the Machiavellian Henry Kissinger in his dealings with Communist China, was the survival of the United States so threatened that the use of such marginal means were necessary? How many American lives did they save because the war was ended by this extreme means? It has been stated that the strategies of Machiavelli show no prejudices for good or evil means. They have a disregard for the principles establishing the power structure.

President Truman may have used Machiavellian principles to support democracy or was too ignorant to understand its implications.[10] Again, taking the premise of this book a step farther; in a Machiavellian-directed conflict that requires all that is available to win the *fight*, we need to continue to ask ourselves if the force applied to achieve victory (Machiavellian or not) on the battlefield includes a relevant termination point and a life cycle focus that brings it through resolution? The answer may lie in the amount of resources a society is willing to apply (from cradle to grave) to the fight.

"WHOLE OF" APPROACH

Talk of integrated civil-military plans and joint campaign plans cannot disguise their lack of reality of well-managed civil efforts, proper resourcing of programs, the stove-piping and lack of basic accountability in most aid efforts, and the near chaos in managing the overall foreign aid effort within the interagency. Moreover, military forces engaged in COIN operations invariably find themselves involved in civic action programs designed to alleviate social, economic, medical, and other problems at the local level.

It is instructive at this point to give the reader a further understanding of what appears to be the buzz phrase for the Obama Administration's first 100 days—"whole of government." As a review from earlier parts of this book, whole of government, or WoG (and whole of nation—WoN or whole of world—WoW construct), is the understanding that conflicts cannot be fought, and more importantly, cannot terminate or come to resolution, without a synchronized effort across all agencies in our government.

Critics argue that the DoD is the most resourced (money, equipment, and manning) agency in our national security apparatus and, therefore, should carry the largest load of the work. This criticism has proven prophetic because in Iraq and Afghanistan the DoD is carrying the burden. My own experience as a Brigade Combat Team Commander in South Baghdad for 15 months from 2007 to 2008 has given me a clear understanding of the shortfalls inherent in agencies other than DoD.

Specifically, my interaction with the embedded Provincial Reconstruction Team (ePRT) and the United States Agency for International Development (USAID) has validated my understanding of the shortfalls of agencies other than DoD. Although all are motivated to perform well, some just are ill equipped, or in some cases, not equipped at all to accomplish the task; and in this includes DoD.

Regardless, the lack of skill sets within my brigade combat team (BCT) didn't prevent the command from finding ways to infuse successes in the local economy. Agricultural, poultry, and fish farming were just a few examples of the potential the populace required in the early stages of success brought in by the OIF surge of 2007 to thwart the insurgents' capacity and capabilities to control the population. In my example, the Southern Belt of Baghdad had been infamously labeled the "Triangle of Death" where the populace was vulnerable to the influence of the enemy *if* the governance couldn't provide—and it couldn't in the early years of the war. Therefore, it was a critical time where there was little room for error as the populace was teetering on edge of either supporting the insurgency or the efforts of the Coalition to support the local government. This is an example where the military clearly needed a WoG approach, and the interagency support was absolutely required.

And although the military was unable to conjure up the skill sets asked to perform the full array of economic and governance talents, the interagency didn't have the resources to employ its expertise. However, in the end, bilingual and bilateral advisers and a robust PRT augmented by a number of military personnel met the task through facilitating the progression of local projects. With the employment of what military and civilian expertise we could find, the mission proved successful; however, it was all done after the fact and not part of the original plan.

WHAT IS NEEDED?

Failing and postconflict states pose one of the greatest national and international security challenges of our day, threatening vulnerable populations, their neighbors, our allies, and ourselves. On August 5, 2004, Secretary Powell announced the creation of the State Department's Office of the Coordinator for Reconstruction and Stabilization (S/CRS) to enhance our nation's institutional capacity to respond to crises involving failing, failed, and postconflict states and complex emergencies.

The development and success of the S/CRS is critical to any postconflict model. The development of the CRS is only one facet presented by the 2008 *Project of National Security Reform* government reform that is center stage in President Obama's first term. It appears that with the implementation of S/CRS, society is now, at the very least, taking on the burden of augmenting the military effort with civilian expertise. With that said, the core mission of S/CRS is to lead, coordinate, and institutionalize U.S. government civilian capacity to prevent or prepare for postconflict situations, and to help stabilize and reconstruct societies in transition from conflict or civil strife, so they can reach a sustainable path toward peace, democracy, and a market economy.[11]

As described by Secretary of State Clinton, today's world is interconnected. Whatever country we are from, we share urgent challenges that transcend borders—climate change, food security, diseases, energy, terrorism, piracy, and, of course, the global economic crisis. Clinton has committed the U.S. State Department to a new diplomacy powered by partnerships, pragmatism, and principle. This commitment is elevating development to its rightful place alongside diplomacy as a key component of our international efforts. The concept is to promote good governance, human rights, and social inclusion so that more people around the world can claim their rightful share in global progress and prosperity.

In a rapidly and continuously changing global environment, failing and postconflict states pose perhaps the greatest challenge to stability in the world. Recognize that in the past, the global community addressed reconstruction and stability issues in an ad hoc fashion, securing and refashioning the necessary tools, strategies, and relationships anew with each crisis. To address this shortfall, in 2004 the office of S/CRS was chartered to build and maintain an "expeditionary, innovative, and interagency" civilian capability to plan, manage, and conduct stabilization operations.

S/CRS is using new tools and seeking new partners to broaden the reach of our diplomacy understanding that 21st century statecraft cannot just be government-to-government; it must be government-to-people and people-to-people. To pursue its agenda, S/CRS engages civil society, women, youth, political activists, and others as we pursue our agenda. In short, the mission of S/CRS is to lead, coordinate, and institutionalize a capable capacity to prevent or prepare to resolve postconflict situations, and to help stabilize and reconstruct societies in transition from conflict or civil strife so they can reach a sustainable path toward peace, democracy, and a market economy. [12]

Of note, S/CRS is seeking constructive solutions specifically in the Middle East, where they have made a major commitment to assist the Palestinian people, and in Iraq, where they are working toward a responsible redeployment and, perhaps more importantly, replacement of U.S. combat forces and taking a new approach to Iran that relies on all the tools of U.S. power, led by diplomacy.[13]

The organization of S/CRS is such that its concept is to provide the U.S. government and as it's implemented, the international community, with a pool of qualified, trained, and ready-to-deploy civilian professionals to support overseas reconstruction and stabilization operations. The organization is structured to plan to prevent and to react to the conflicts. Although its success is yet to be seen, conceptually S/CRS provides the interagency exactly what is missing; skill sets to assist in reconstruction and a planning team to imbed early on to devise how they can best support conflict resolution once war termination occurs (see Figure 5.1). [14]

THE FUTURE OF FOREIGN ASSISTANCE

Even if the agencies in the formal government worked well together, and more importantly, are resourced to do so, we are only partially optimizing the skill sets available. To meet the premise offered by this text, those skill sets required for stabilization and reconstruction have to be established to properly plan for war termination and conflict resolution to meet a desired endstate before the fight begins. Reaching a new consensus internationally for development is critical in this effort. A new consensus among the larger community of development experts and partners, corporations and contractors, entrepreneurs, nongovernmental organizations (NGOs), foundations, Congress, and the federal government as a whole would bring to the fight the means to achieve this endstate. More importantly, knowing these organizations are resourced and available and

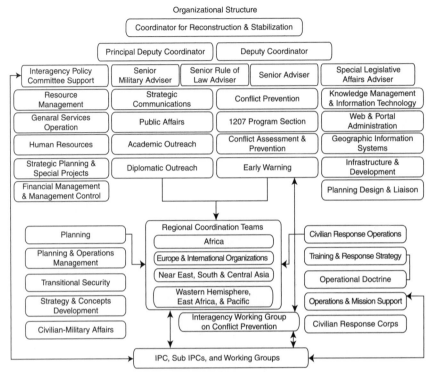

Figure 5.1

motivated could be the genesis for a better synchronized plan prior to the fight.

This new consensus would be the development equivalent of a Galilean revolution: one that puts the host-country—its priorities, capacities, norms, and local design—at the center of our collective thought and action around which the rest of our organizations would revolve, be rewarded, and exist to serve.

Why is it that we are all rewarded for excelling as individual organizations when in fact we must learn to operate as a cohesive, coherent, coordinated whole? Why are we all passionately devoted to our organizations, disciplines, agencies, regions, or sectors when host governments say what they want from the development community, the private sector, and the U.S. government is coordination of our efforts and direction? "Rather than operating as a thousand forces for good—which at times leads to a thousand points of contention—foreign assistance funding and execution would be far less supply-side oriented, and much more demand-driven."[15]

What is not well-known, and arguably more shortsighted, was the gutting of the U.S. ability to engage, assist, and communicate with other parts of the world—the *soft power*, which has been so important throughout the

Cold War. The State Department froze the hiring of Foreign Service officers for a period of time. USAID saw deep staff cuts—its permanent staff dropping from a high of 15,000 during the Vietnam War to about 3,000 in the 1990s. And the U.S. Information Agency (USIA) was abolished as an independent entity, split into pieces, and many of its capabilities folded into a small corner of the State Department.[16]

Although this solution arguably should come from the international community, there is no doubt that the United States leads the way in developing this effort. There are two proposed approaches to fixing the problem: integrating USAID even more completely into the State Department, with its head reporting directly to the secretary of state, or granting it significantly more independence, either as its own cabinet-level department or as a strong autonomous agency. The first option would make things even worse than they are now because there is a strong conflict with USAID and the State Department.

In a full merger with the State Department, USAID would lose its development mission altogether, as that mission would continually lose out to the State Department's more traditional diplomatic priorities. The right approach is to find some way of restoring USAID's autonomy and vitality. Unfortunately, instead of attempting to bolster resources to make the interagency a more robust, capable effort, most of what is going on is stove-piping concerns as to who is in charge. The real question is whether USAID should be an independent agency reporting to the secretary of state or a new cabinet department. Both routes have advantages and disadvantages, but either would be preferable to the current setup.

There is no doubt that a cabinet-level department would give USAID much greater stature and allow it to influence policy on trade, investment, and the environment while improving existing assistance programs. This approach is the predominant model used in wealthy donor countries. The United Kingdom moved in this direction in the mid-1990s. The United Kingdom's Department for International Development has used its perch to achieve greater influence on development matters throughout the British government by helping to shape trade, finance, and environmental policy at the cabinet level. As a result, the Department for International Development has become the most prominent government aid agency in the world, even though London spends far less on aid than Washington. The chief argument against a cabinet-level development department in the United States is that the secretary of state needs to have policy involvement and oversight when it comes to foreign aid. Moreover, the secretary of state is always going to be a more powerful member of the cabinet than a development leader.

USAID often relies on the active support of the secretary of state to get the funding and legislation it needs to carry out its mission. For these reasons, many observers believe that a strong agency reporting to the secretary of state would be preferable. As of today, USAID is subordinated to the Department of State (DoS) and although the hierarchy is tenuous, there are several policies that must be implemented to strengthen U.S. development capabilities.

First, the new USAID must have budgetary independence, and its operating account, which pays for buildings, salaries, and technology, should be dramatically increased to boost the size of the permanent staff, invest in training, and increase the agency's technical expertise. This will enable the new USAID to reopen missions that were permanently closed and to staff them adequately. Second, the head of the new USAID should be a statutory member of the National Security Council and serve as part of the president's international economic advisory team on the National Economic Council. There are compelling security and macroeconomic arguments for foreign aid.

As Paul Collier's acclaimed book, *The Bottom Billion*, demonstrates, countries with high poverty rates descend into civil war far more often than more prosperous nations. These conflicts kill thousands and destroy the political and economic institutions of the states in which they occur, leaving the international community to pick up the enormous tab for rescue, relief, and reconstruction.[17] The goal for organizations like S/CRS is to get ahead of this conundrum. Proper planning and resource allocation can prevent failed state status and the potential for civil war. In fact, with the implementation of the functions associated S/CRS and its habitually tagged resources like USAID, conflict resolution outcomes could turn positive. Conflict resolution and resource allocation is more complex than just *wanting* to do well. It is a complex challenge that requires graduate-level apprehension.

Development success is closely related to investment, trade, and finance policies; U.S. policy and developing-country policies on these matters are as important as the volume of foreign assistance. U.S. agricultural subsidies, trade protectionism, and subsidies for ethanol all hurt poor countries by distorting food markets. Yet within the U.S. government, decisions concerning international trade and finance are all too often made without any regard for reducing poverty or stimulating economic growth in poor countries.

Making USAID an integral part of the interagency process allows it to influence policy making and take direction from the DoS, the Pentagon, and other agencies on matters involving foreign policy and national security. But to garner the required resources, the new USAID will need a new congressional mandate. The Foreign Assistance Act of 1961, which has not been amended in any meaningful way since 1985, is a Cold War artifact that has become obsolete.

Organizations like the Millennium Challenge Corporation (MCC) can provide assistance to provide credible evidence as to the need of monetary support for many of the interagency departments. Their purpose, as stated by the MCC's home page on the Web as an innovative and independent

U.S. foreign aid agency that is helping lead the fight against global poverty, could serve as the basis for new legislation.[18] The MCC is an innovative and independent U.S. foreign aid agency. It was created by Congress in 2004 with strong bipartisan support. In short, MCC is changing the conversation, as advertised, on how best to deliver smart U.S. foreign assistance by focusing on good policies, country ownership, and results. Before a country can be eligible to receive assistance, MCCs board examines its performance on independent and transparent policy indicators and selects campaign-eligible countries based on policy performance.

As stated for eligibility criteria, the MCC uses 17 indicators in three broad development categories—"ruling justly, economic freedom, and investing in people"—to determine a nation's eligibility to receive development aid. A new congressional mandate would make the executive branch accountable for results and provide a new framework for legislators who wish to earmark funds for specific purposes.[19]

As of 2009, the United States started using a new term, security sector assistance (SSA) to propose a shared responsibility to pool resources that is responsive to crisis needs as well as designed to mitigate the risk of U.S. military action. Building on a UK concept, this approach envisions pooled funding mechanisms for (1) security capacity-building, (2) stabilization, and (3) conflict prevention.[20] Each department would seek and potentially secure funds within its budget to contribute to the funding pools. In turn, each department would be able to add funds to the pool to meet a departmental imperative although the use of these funds would be subject to the dual-key approved requirements (see Figure 5.2).

In the near term, however, such an approach is best suited to succeed the pioneering Sections 1206[21] and 1207[22] programs, which have notably incentivized interagency collaboration.

The United States has a major interest in providing security assistance that is responsive to crisis needs as well as designed to mitigate the risk of U.S. military action. Programs that currently capture these priorities— sections 1206 and 1207—can fairly claim some notable successes in the past several years. On the other hand, they have stirred debate over U.S. government roles and missions. The ongoing jurisdictionally-based debates about the resourcing of conflict support have levied a major tax in terms of adjudicating disputes at high levels in the Executive Branch and with Congress.

The UK's pooled funding model allows its ministries, to include the Ministry of Defense, Foreign Ministry and Department for International Development, to contribute funding to a joint pool designed to incentivize integrated approaches to cross-cutting issues. The pool is overseen by an integrated tri-departmental staff and tri-departmental approval structure. Originally, each agency contributed funds from within its budget. Now that the fund is more mature, it competes for funding directly from the UK treasury, although agencies may still add to the pool with additional

	Security Capacity Pool	Stabilization Pool	Conflict Prevention Pool
Field-Level Proposal Development	Projects endorsed by the Chief of Mission and Combatant Commander	Project endorsed by the Chief of Mission, USAID Mission Director, Combatant Commander	Project endorsed by the Chief of Mission, USAID Mission Director, and Combatant Commander
DC-based Staff Support	Near-term: Each department or agency provides staff support to Senior Steering Group representative. Long-term: Single, collocated staff of interagency detailees. Process Secretariat: DoD	Near-term: Each department or agency provides staff support to Senior Steering Group representative. Long-term: Single, collocated staff of interagency detailees. Process Secretariat: State	Near-term: Each department or agency provides staff support to Senior Steering Group representative. Long-term: Single, collocated staff of interagency detailees. Process Secretariat: State
Senior Steering Group	2-star level representatives from OSD(P), Joint Staff, and State/PM. By-laws would forward disagreements for 4-star level adjudication after 30 days (USDP, VCICS, T)	2-star level representatives from OSD(P), Joint Staff, S/CRS, and USAID. By-laws would forward disagreements for 4-star level adjudication after 30 days (USDP, VCICS, T, USAID Administrator)	2-star level representatives from OSD(P), Joint Staff, S/CRS, and USAID. By-laws would forward disagreements for 4-star level adjudication after 30 days (USDP, VCICS, T, USAID Administrator)
Final Approval Authority	Dual-Key (SecDef & SecState)	Dual-Key (SecDef & SecState)	Dual-Key (SecDef & SecState)

Figure 5.2

funds to meet a departmental priority. Because such an interagency appropriation would be difficult given the U.S. congressional structure, a more likely mechanism is for DoD, DoS, and USAID to each secure funding for their contribution to the pool. Given the imbalance of DoD budget to that of other agencies, especially DoS and USAID, it would be a challenge, but not a reason NOT to pursue (see Figure 2.1).

As stated, the three funding mechanisms are (1) security capacity-building, (2) stabilization, and (3) conflict prevention. Each department would seek funding within its budget to contribute to the funding pools. Each pool would operate with joint formulation requirements in the field and dual-key concurrence in Washington, DC. Legislation would endow these funds with inherent authority to achieve their purposes. Each department would be able to add funds to the pool to meet a departmental imperative, although the use of these funds would be subject to the dual-key approval requirements.

Given the U.S. emphasis on building partner-nation security capacity and the relative expense of such activities, creating a separate pool for each activity is key. This approach also comports with a lesson offered in interviews with UK officials that encumbering a single pool with too many purposes can create a structure that is "transaction heavy" (i.e., competing objectives create sclerotic processes that hinder effectiveness).[23] Separate stabilization and conflict prevention pools would ensure that preventative measures are not displaced by current operations. All three pools would be targeted to fund programs with a clear security nexus. Assistance that primarily supports traditional defense policy, foreign policy, or development objectives would still be funded separately by DoD, DoS, and USAID under their existing authorities.

In theory, a shared responsibility pooled resource option approach could be applied to any activity that would benefit from joint planning and programming. If successful, a pooled approach could provide a model for other initiatives. As stated, under this construct the security capacity building pool would assume the role currently provided by Section 1206. Section 1206 provides training and equipment to partners for counterterrorism operations or for stability operations in which U.S. Forces are a participant, including Afghanistan.

It has had some notable successes, including providing training and equipment for the Lebanese Armed Forces during their 2006 conflict and building maritime domain awareness, information-sharing, and increasing interdiction capacity in Indonesia, the Philippines, and Malaysia that has contributed to greater stability in the Malacca Straits. The security capacity building pool would include training and equipment for combating terrorism and stability operations, and could be expanded to other near-term operational priorities. Institutionalizing this capacity would ensure our Chiefs of Mission and Combatant Commanders have the tools they need at their disposal while pooling resources from both DoD and DoS and would reflect the requirement to meet unplanned train-and-equip requirements for foreign policy needs as well as for operational military requirements.

Similarly, separate conflict prevention and stabilization pools would perform the stabilization purposes originally intended for Section 1207 as well as preventative measures to reduce the likelihood of crises and conflict. The stabilization pool would address stabilization requirements such as assistance to Georgia for both its police and internally displaced persons following the 2008 invasion of Russia. The conflict prevention pool would be targeted at preventative measures intended to reduce the risk of crises as well as actions designed to seize windows of opportunity such as extending the government of Colombia's writ after unanticipated successes against anti-government insurgents in the La Macarena region in 2007 and 2008.

The management of shared responsibility pooled resources clearly will be difficult; especially based on historical struggles trying to get 1206 and 1207 funds defined. Nonetheless, 1206 and 1207 cases can be the catalyst for interagency collaboration. This has been done with programs often facilitating joint efforts by combatant commands (COCOMs) and embassy staffs. For each program, joint DoS-DoD formulation of projects is required by law. In the case of Section 1206, the policy requirement for both the Chief of Mission and Combatant Commander to endorse projects prior to review in Washington, DC, further incentivized collaboration. The pooled approach would retain the concept of joint formulation and would include USAID in projects funded by the conflict prevention and stabilization pools.

For each funding pool, a senior steering group comprising the Deputy Assistant Secretary-level representatives from each organization, supported by an interagency staff, would review proposals and forward recommendations to the Secretaries for final approval. The senior steering groups would also establish bylaws requiring Under Secretary-level adjudication of disagreements after 30 days. Each organization would serve as the secretariat for one of the pools and perform the administrative functions required for its operation.

In defining what authority and funding would be required, this approach would require:

- Legislation establishing these funding pools in the Treasury of the United States (as outlined in Figure 5.2).
- Each department would need to seek authority to provide funding to the pool as well as its own appropriation for this purpose. This funding could be drawn from existing accounts or identified as a separate funding requirement with a dedicated appropriation each year. The request for authority would likely also include a mechanism for each department to be able to add to the pool if a departmental priority needs to be addressed in the near term.

The division of funding needs further consideration. To offset DoD monopolization of recourses "matching shares" requirements with DoS and DoD splitting the security capacity building pool, and for the conflict prevention and stabilization pools, DoD providing half and DoS and USAID collectively providing half. An alternative might be for each department to fund each pool relative to the priority it places on its particular function. Under this rubric, DoD might contribute more to the security capacity-building pool given the relative importance of training and equipping partners for COCOM security objectives while contributing less to support primarily-civilian preventive efforts. The ability of each department to secure appropriations will be critical for long-term success, and will likely require burden-sharing negotiations between DoS, USAID, DoD, and their respective committees.

Oversight will be the difficult part. Using the 1206 and 1207 model is cumbersome at best. Instead, the creation of a select committee in both the House and Senate to oversee these funds is an option. For example, the House and Senate could create a Select Committee on Security Capacity-Building, Stabilization, and Conflict Prevention. Most radically, Congress could establish a new title of U.S. Code, separate from DoD's Title 10 or DoS's Title 22, under the oversight of these Select Committees. A new "Title 51" would codify this approach in law and demonstrate that these programs are cross-cutting and not appropriately captured within any single committee's jurisdiction or section of U.S. Code.[24]

CHAPTER 6

The Nesting of Goals and Objectives to Achieve an End

It's a triumph. What thoroughness! What realism! Knew when to stop, too—
didn't cut the pages. But what do you want? What do you expect?
— F. Scott Fitzgerald, *The Great Gatsby*

This is a chapter that discusses how the United States can be better organized to support the natural struggle at the policy level for defining strategic goals.[1] It describes the cascading effect of not achieving a standard for ending wars. The United States must figure out how to define strategic exit strategies, or else. This chapter defines what that "or else" is as it relates to the military, the national reputation, and its resources as well as the impact on the global community. The majority of this chapter gives the reader an overarching feel for the long-term impact of not achieving "victory" based on not defining goals, objectives, endstates, and war termination strategy.

The key is to be able to clearly define both the political conditions and the situation that one envisions existing when both the conflict and dispute are over.[2] This is the leader's vision; a key part of establishing goals, objectives, endstates, and a termination strategy. Growing an objectives tree, so to speak, is what we are attempting at this point: Understanding the nesting of goals and objectives with a desired endstate by:

- Defining the goals
- Devising an objectives tree to show how they are linked

The nesting of goals and objectives in the form of objectives trees is a systems engineering approach that I used in curriculum while an Assistant Professor at the U.S. Military Academy at West Point (see Figure 6.1 for an example of defining nested goals, objectives, and endstates). The process of developing an objectives tree to help define a baseline set of measured objectives that lead to definitive endstates is a tool that I suggest all problem solvers and strategists use when confronted with a multidisciplinary, complex problem set.

The first step in developing an objectives tree is to define all issues as a set of problems, for instance, from our example, "expelling the Iraqis from Kuwait," and then inverting these problems to make them into objectives. An objectives tree is a tool used for clarifying the objectives (and associated goals and constraints) of a project, or in our case, in military or policy terminology, a mission. It starts with a set of broad goals as expressed by the mission statement and expands them into as many subgoals as possible. It leads to a treelike diagram of the hierarchy of goals and their interconnectivity.

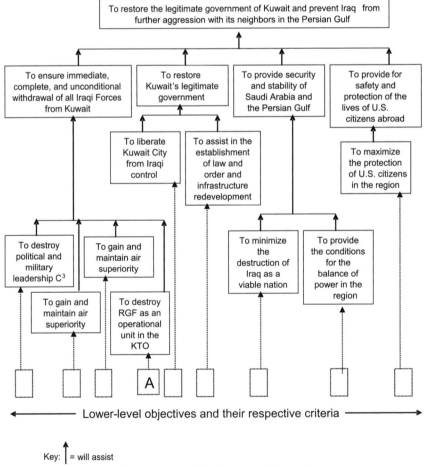

Figure 6.1 Goals and Objectives Tree for Operation Desert Storm

Although this may not be a novel idea, it forces strategic planners to exhaustively define an endstate, its goals, and the eventual means to achieve each. The intention, as it pertains to this text, is that, if objectives and goals are thoroughly defined, the strategist will have little choice but to determine the war termination and conflict resolution specifics.

The world is changing and now is the time to not be overcome by swift changes occurring because of globalization, technological revolutions, and the flattening of the world economically and socially. In the complex and increasingly interconnected world of the 21st century, multilateralism will not be an option, it will be necessary.[3]

Many believe that this reductive type reasoning flies in the face of the holistic, whole of government (WoG) systems approach that I have championed in this text. However, the objectives tree construct is not merely an attempt to explain the complex effect through a simple cause. In fact, its display conveys in visual form a need to build in criteria for success to achieve the effects-based objective of meeting each goal. Without the identification of criteria, each objective runs the risk of being affected by mission creep or consequence management needs; both of which we decided earlier were a result of poor planning.

A holistic approach denotes the "circling of the wagons" for resources and organizational efforts. The "Whole of" concept as described earlier encompasses the holistic approach. To strengthen the civil instrument of a national power to leverage the full potential of governments like the U.S. government, a WoG capabilities check is necessary to manage national security issues that are by nature complex, dynamic, and of international concern. "Even in Iraq in its reconciliation effort to identify reconcilable and those nonreconcilable requires a holistic approach."[4]

In short, there is no single agency for every country and no solution that can homogeneously support the strategy that will endure over time to solve these challenges. The full array of tools at a nation's disposal include diplomatic, economic, military, political, legal, and cultural; picking the right tools for each solution is difficult but critical when defining what's needed for conflict resolution. Some doctrine experts such as Vego conclude that,

> The impetus toward adopting an effects-based approach came in the aftermath of the Vietnam War (1965–1975). Then, the U.S. military emphasized the need to link objectives at all levels of war—from the national political level to the tactical—in a logical and causal chain. In their interpretation, this outcome-based or strategy-to-task approach became the basis for joint planning.[5]

FROM BLOGGERS TO WOGGERS

WoG, nation (WoN), or world (WoW) mass collaboration is based on the theory that collective actions' activity through large numbers of participants and providers working autonomously, but with common purpose toward modular aspects of a single issue, can dynamically unleash greater innovation, resourcefulness, and situational awareness. Collaboration is managed not through control or directives but based on the collective consciousness of self-enforced activity of those choosing to participate and mediated by the shared understanding of an issue or threat. This, therefore, is not about amassing "cooperation" because there are no negotiation and directives motivating consensus.

The heart of the collaboration is based on a common "hearts and minds" purpose defined by a common issue that is communicated in global reference through all forms of mass media down to the lowest levels of personal interaction. Such issues could be infectious diseases, global terrorism, peacekeeping/conflict prevention, education, natural disaster prevention or mitigation, human rights and rules of social law, illegal drugs, etc. Without the common purpose expressed effectively, mass contributions will never amount to anything along a focused cause to "direct" efforts.

The "Whole of" effort is developed to better defend and/or attack contested zones of influence that reside within the needs of populations, host-nations, and the intervening forces that vie for positional advantage, desired effects, and outcomes. The framework for mobilizing zone of influence support and defining a message that will develop communities of purpose and a means to manage that effort will take nothing less than their own grand scale of irregular warfare methods to influence and sustain the common purpose against an adversary or threat including:

- Concept of the operation/establishment of the purpose
- Intelligence/information support
- Recruitment of localized resources and surrogates
- Modern social-network variance of command, control, and communications (C3)
- Civil-military-government operations
- Information/influence operations
- Cross-cultural leadership skills
- Psychological operations
- Counterinsurgency "oil-spot" approaches[6]

We have to, at the very least, understand globalization. Barnett states,

> Show me where globalization is thick with network connectivity, financial transactions, liberal media flows, and collective security, and I will show you regions featuring stable governments, rising standards of living and more deaths by suicide than by murder. These parts of the world I call the Functioning . . . show me where globalization is thinning or just plain absent, and I will show you regions plagued by politically repressive

regimes, widespread poverty and disease, routine mass murder, and—most important, the chronic conflicts that incubate the next generation of global terrorists.[7]

Postconflict threats are real. Stumbling into an insurgency such as what happened in Iraq in late 2003 is a result of no postconflict and war termination strategy. Other potential challenges include rising state powers, rogue proliferators, and nonstate actors that include terrorists, transnational criminal organizations, and other assorted entrepreneurs of violence.

CHAPTER 7

The Art of Ending War

The first, the supreme, the most-far-reaching act of judgment . . . is to establish . . . the kind of war on which they are embarking; neither mistaking it for, nor trying to turn it into, something that is alien to its nature.
—Carl von Clausewitz

Thus far, I have presented key terms, definitions, vignettes, and examples of war termination successes and failures; a whole of government, nation, and world overview; and a review of our national security strategy and the nesting of the military strategy. Readers looking for an unveiling of a grand methodology that is different from what exists in the form of mission planning and strategy development will be disappointed and perhaps at the same time, relieved, at this point. Disappointed because there is a natural reaction to expect the discovery of a dogmatic panacea unfounded in our national security planning process; relieved, however, because the methodology already exists; we just need to follow it.

What the United States lacks as it continues to decide to deploy troops to a fight is the discipline to identify an endgame strategy and the hard decisions to resource it. The resourcing is a key—if not decisive—ingredient. Dollars or, more accurately, money is essential when devising plans for reconstruction and stabilization. Time is a resource in this case and, as always, time is a zero sum gain, especially in a crisis mode. The lack of time is why the development of a war termination strategy is often overlooked. The attitude that we will figure it out on the fly and focus on defeating the enemy is prevalent.

Although the armed forces have the predominant role on the battlefield, the combatant command is but one actor among several during conflict termination. The process requires interagency (and often coalitionwide) cooperation to deal with the diverse political, economic, humanitarian, and military issues. Rarely will conflict be resolved through the finality of unconditional surrender; limited war is the rule, and total war the exception. Accordingly, the United States must have the benefit of a variety of perspectives and expertise as it adjusts from war to a new and, hopefully, more favorable peace.

Capitalizing on what has been explained thus far and extending an understanding of nesting goals and objectives to achieve an endstate as explained in Chapter 6, we can now establish a methodical process for achieving an endstate. Keeping in mind that this is not a linear process, the first step is to fully define the endstate.

After verifying initial objectives, the first task would be to recommend the desired outcome of a conflict. In some cases, this might mean taking the initiative in planning. Crises develop quickly and unexpectedly, and the national security team may not have time to fully define all the goals of an operation. Restore Hope was a case in point. A tactical planning staff had to assist the chain of command with desired endstate planning, albeit with less than optimal results.

GATHERING THE TOOLS

An extensive interagency planning team imbedded into the crisis action or deliberate planning process would have lifted this additional burden from the military and given endstate definition the attention it demands. Such a process would encourage senior leaders to conduct serious deliberations on the subject and allow the rest of the planning staff to focus on deployment and initial employment of forces. The military decision-making process is an existing planning tool that can translate easily to the strategic planning process to include the interagency involvement. The addition of this multifunctional team to the military planners can allow for a focused effort on part of the operation that includes governance, economics, diplomacy, and other areas in which the military lacks expertise. Imbedding the team to cover military and civilian capacity-building, as noted in the recent Operation Iraqi Freedom (OIF) and Operation Enduring Freedom (OEF) conflicts, provides a focused effort on building the indigenous force that will (and has to) eventually replace the fighting force whether or not the initial effort is an invasion force or advisory assistance military effort.

The next step is defining military transition conditions. The transition focus may be the most difficult part to devise, and because it is time consuming and complex, most planning teams tend to forgo the mental gymnastics involved to commit energy to defining war termination and conflict resolution phases of a fight and spend the majority of their allotted planning time (in a crisis mode, time could be severely limited) on the employment of lethal forces to destroy the enemy; the exciting part of any fight for a military or strategic planner.

This part of the fight is the military's comfort zone and, quite frankly, the most compelling crisis that needs to be solved in the early stages of any fight. Regardless, by developing a well-thought-out objectives tree or by using some other visual tool to help define the endstate and by using that endstate as a non-negotiable goal, the chance is that the decision-making process will include war termination criteria, thereby ensuring that with a deliberate effort, a conflict resolution plan is much more likely.

After achieving consensus on the endstate, the team would assist in defining the military conditions that will lead to a successful transition to diplomatic leadership. These conditions would become military objectives for combatant commanders. In conjunction with military planners, the multifunctional planning cell could advise on the appropriate ways and means to achieve these objectives. Its mission would be to incorporate and synchronize all key dimensions in the plan.

Sequencing is the natural next part of this process. Favorable transition conditions will take time to evolve. Thus the cell's next task is to develop a sequenced path to the military transition state. This may be a series of phases where the generation of specific circumstances may signal the end of one phase and the beginning of the next. Bruce Clarke developed a synchronization matrix that could be used for phased transition state planning (see Figure 7.1). With this tool, Clarke emphasizes that "if we have done our planning properly we have the cease-fire documents or whatever mechanism we have elected to use to terminate hostilities ready long before hostilities have ended."[1]

According to Clarke, if we have synchronized each action that we take, our defensive actions have anticipated the opponent's actions and thus denied him the initiative; then, we will achieve some form of settlement. The synchronization can be shown graphically in a series of boxes with the elements of power (hard and soft) listed on the vertical axis and the phases and subphases of the conflict listed on the horizontal axis and the probable/possible opponent's reaction also. This is Clarke's depiction of how conflict unfolds and his way of reducing the Clausewitzian friction or fog of war.

To reduce this friction, we talk about the synchronization of activities so as to bring the requisite power to bear at the decisive time and place. If done properly, it also reduces the complex into a series of smaller events and ultimately allows the planner to see the need to devise a war termination and conflict resolution strategy to synchronize all the resources and events in the process. A similar matrix can be built for the individual components of economic, political, or military power within the agencies primarily responsible for execution of a portion of the synchronization, thereby enhancing the utility of that element of power and optimizing all skill sets required to bring the conflict to resolution.[2]

Figure 7.1   Strategic Synchronization Matrix

Strategic Vision	Dispute	Prehostilities	Hostilities	Posthostilities	Settlement
Time/Phase					
Political					
UN Support					
Communication					
Alliances					
Economic					
Tariffs					
Sanctions					
Most Favored Nation Aid					
Socio-psychological					
Cultural/Ideological					
Legal—World Court					
Military					
Deterrence					
Coalition Warfare					
Security Assistance					

This process shows the planned status of variables such as command and control, security, economy, and diplomacy by operational phase. Granted this is a linear methodology and, although its visual design portrays a sequence, the display of each phase is provided linearly in fashion with the full expectation that many of the activities are done in parallel. With that understanding, the operation moves to the next phase when a variable meets the tripwire definition described in the matrix.

The variables used must be measurable benchmarks that help those executing the plan know when to employ resources against the "next" phase. One of the major challenges in the process is that the planners who develop and understand what tends to be a complex, integrated plan with multiple nuances need to know that a slight misunderstanding of when to engage a volatile indigenous government, for instance, would cause great chaos in reconstruction and stabilization operations.

Unfortunately, the individuals who are the planners are not the same as those who execute the plan. This causes an even more complex and burdensome challenge to an already difficult task. Therefore, the conveyance of objectives and the measurable criteria have to be precise and captured in a tool, such as an objectives tree or Clarke's modeled synchronization matrix (or both) during the planning process and transferable in understandable terms for any organization or set of individuals to execute.

Given the understanding to translate developing and executing such a plan, the synchronization matrix (as well as the objectives tree) is something that can assist in translating what was planned into what has to be done. Therefore, this tool could be tailored to any crisis. In the end, although most planning cells that develop complex transition plans aren't the same ones that execute them, there is no reason why organizations cannot maintain this element in place, especially after recognizing the criticality of its purpose.

When circumstances favor transition, the cell would advise how to maintain this preferred state to continue progress toward the next phase. Ideally, when all transition conditions are met, combatant commanders are ready to hand off leadership to the diplomats, and these planning/executable cells could facilitate this transition as long as the right skills sets (those with diplomatic experience, for example) are available.

An old military adage states: No plan survives contact with the enemy. Political aims may change, the desired endstate could be modified, and conditions that lead to success may vary. Because objectives, endstates, and strategy are a continuum, team members would have an important monitoring and assessment role. The planning/execution cell should conduct a rolling net assessment, taking full account of the economic, social, psychological, and diplomatic aspects of existing circumstances.

It is clear that team members must be integrated into all available theater informational resources to accomplish this vital task. The cell would advise combatant commanders on ways to calibrate objectives and refine strategy. What may appear to be a perfect and exacting transition plan with resources identified, governance assistance teams trained and ready, economic infusion plans established, reconciliation efforts measured on the risks associated with each and indigenous security forces transition plans in place will, as anyone who has been involved in conflict knows, all change as the structure evolves. What is critical and, as the premise of this text defines, is that *a* plan is in place for it is easier (by an order of magnitude when it come to combat evolving into reconstruction and stabilization) to adjust an existing plan than to develop one in total while the fight ensues.

As components of strategy change, regional commanders could provide higher quality feedback to national leaders on the implications of modifying strategy if there is a plan in place, and most importantly, with the resources allocated (usually the bottleneck in executing the plan). Senior leaders are more apt to adjust to what is already in place than to attempt to cut through bureaucracies like congressional purse strings to get a plan approved after the bullets have already started to fly and in the nth hour of the fight.

To mitigate the challenge of having to change midstream, the planners habitually develop contingencies, branches, and sequels in an attempt to war game all possible adjustments based on predictive analysis and historical heuristics taken from past conflicts. Similarly, the war termination planning cell must develop offshoots and follow-on activities that would lead to winning the peace.

A hand-off to diplomatic leaders is a critical transition. The addition of functional experts helps to coordinate with their counterparts from the country teams to ensure a smooth changeover from the military lead to the civilian lead and, when appropriate, finally to the indigenous government. The timing is critical because new governments, especially those replacing dictatorships, oftentimes take a great period of time to form and take root; the new Iraqi government of 2010 is a perfect example of this challenge.

During peacetime engagements, the cell should be directly involved in strategic planning as well as political-military coordination and theater engagement strategy. The combatant commanders could also use the team to strengthen ties with other government agencies. Transition planning exercises could be conducted in conjunction with major joint operations to provide the cell experience in transition state planning. Team members could assist the joint task force (JTF) plans cell as crises erupt and combatant commanders form task forces. Functional experts from the planning cell would be available to act as liaisons with other agencies. Under certain circumstances, such as the employment of a sizable task force for a long duration, it might be wise to stand up an additional planning cell for the JTF commander.[3]

The challenge for military leaders is their habitual difficulty relinquishing command and even control of operations especially after they have been in command during combat operations and lethal forms of combat. This is not a power struggle, but rather a natural tendency of retaining control of an operation environment when potentially lives are threatened and/or have been lost; an emotional attachment to the situation takes root.

Although conflict termination typically generates a complex mixture of policy, economic, and humanitarian issues as well as military concerns, policy matters tend to predominate—particularly with limited war. This is the case, of course, because war is conducted in pursuit of political goals— goals that ought to be within reach at the close of a successful military campaign. Accordingly, if we want to maximize our chances of achieving more than battlefield success, we must have a senior representative from the Department of State and/or National Security Council with the combatant commander during peace talks and in-theater well in advance of the war's termination. An interagency approach best preserves the nation's diverse interests and permits more effective exploitation of U.S. battlefield success.[4] One way to mitigate this anxiety and friction associated with passing the mantle from military to civilian control is to establish from the beginning a mix of staff of military personnel and diplomats.

On the flip side of this issue, U.S. leaders must avoid the temptation to rush into the cease-fire process—to "cut and run" after the battlefield contest concludes. Although strategists would argue that joint publications should clearly remind us of the fog and friction inherent in conflict and of the dangers such disorder brings to war termination, having a definitive plan built into place would all but prevent the hasty withdrawal under fire potential (a la Somalia). No matter how much technological progress a force may achieve, the battlefield will remain a partially shrouded, complex, and confusing environment. One cannot attain a precise picture of the military situation. Moreover, the further removed one is from the conflict, the less complete is one's comprehension of events.[5] U.S. leaders, therefore, must resist the temptation to rush and adjust the decision-making process on war termination and allow the plan and associated relevant facts to develop more fully during the interagency process. Let the next Saddam sit and worry while the U.S. holds his territory, consults with our coalition partners, and patiently explores its options.

Whether planned or not, the war termination and conflict resolution process is demanding. Why is termination strategy so extraordinarily demanding? Is it merely the fog of war? Colin Gray suggests that this particular effort is so challenging because it serves as a bridging function between two dissimilar elements—war and politics. The termination of conflict, as it directly connects these two elements and does so in the course of a transition from one to the other, is at the very heart of this demanding strategic process. We must, therefore, reverse the traditional U.S. approach of divorcing war and politics.[6] In other words, contributing beyond the battlefield to "a better state of peace" requires that the nation's senior war fighters use their limited resources to develop a strategic vision. Capturing and helping implement such a vision requires a combatant commander to spend less time thinking about what his or her forces are fighting against and more time understanding what our nation is fighting for.

SHAPING CURRENT METHODOLOGY

Knowing "why" we went to war can lead to understanding of what the outcome will be and how it can be shaped. Overthrowing a tyrannical government will have the "revenge" condition that can't be ignored. Those that supported the government will want to take their vengeance out on the *occupiers* and those host-nation elements that "support" the occupiers. Not planning for this prior leads to failures as seen in OIF where the insurgents all but snuck up on the Coalition. We also have to realize that if we don't protect those who supported the ousting of the tyrannical leader in the first place, it can result in failure as seen as the Shi'a uprising in the southern part of Iraq after Desert Storm. The United States ignored them and in the early years of OIF contributed in causing a failure to bring peace and prosperity to the newly "liberated" nation of Iraq. But understanding why we fight goes beyond just the recognition of ridding the world of yet another despot. Asking the question "why we fight" should trigger the concern, and of what happens after the bullets stop flying.

To recognize a model or methodology we need to first define some rules. In the same way that I have discussed defining a war termination strategy as early as possible in a campaign, I want to define, in the first stages of this process, what I am attempting to achieve; my exit strategy in this process is having the reader understand the methodology. Recapping what has been discussed thus far in this text:

- Understanding the nesting of goals and objectives with a desired endstate:
  - Defining the goals
  - Devising an objectives tree to show how they are linked.

- Organizing resources and organizations (agencies, nongovernmental organizations, and international organizations) that are needed to accomplish the task in a whole of government (WoG) approach:
  - Listing out tasks drawn from objectives and matching resources against those tasks.

- Developing a timeline showing when to introduce each resource and what its command relationship with other organizations (unity of effort [UoE] and unity of command [UoC] discussions).

- Matching categories for which war termination and conflict resolution correlate to devise a proposed solution to an exit strategy:

Before defining any construct we must first recognize the principles associated with that methodology. In keeping with a need to make lists and chart compartmentalized categories, there are six conflict termination principles when applied to operation art that enhance the chances of successful war termination and conflict resolution—understanding these in the planning process can add to the potentiality of successfully achieving an endstate:[7]

1. Articulate political and military strategic objectives in terms of clearly defined endstates.
2. Apply national UoE and WoG, Whole of Nation (WoN), and Whole of World (WoW) approach including critical government agencies in the war termination and conflict resolution planning process. Ensure a formal relationship is established and maintained for UoE or UoC.
3. Plan for war termination and conflict resolution at the beginning of the planning process.
4. Link the strategic and operational goals to accommodate war termination and conflict resolution through operational design (use of sequencing, branches, and sequels).
5. Synchronize and coordinate posthostility operations between military organizations and civilian agencies. Ensure the appropriate representation from all the key players in the war termination and conflict resolution phases are resident throughout the process (cradle-to-grave approach).
6. Insist on maintaining key planners on the executable staff so that ambiguities of understanding the complex plan are mitigated.

Additionally, we should review in detail the following as categories in specificity for conflict resolution:

- Security forces
- Budget requirements or resources needed and available
- Government needs
- Internal issues to include the status of governance and economic capacities
- Organization of resources and organizations (agencies, nongovernmental organizations, and international organizations) that are needed to accomplish the task in a WoG approach by:
  - Listing out tasks drawn from objectives and matching resources against those tasks.

- Developing a timeline showing when to introduce each resource and what its command relationship is with other organizations (UoE and UoC discussions).

We cannot ignore the fact that people who are put in charge in a new type of government like democracy tend to want to make money like the ones who were in office before them. So corruption, although an ancillary effect, is a virus that can bring any well-defined postconflict resolution plan to a halt.

As one may recall from previous chapters, established compartmentalized categories for terminating war is critical in infusing likely outcomes to the fight. Chapter 1 defines each of these following categories. The planning cell should consider the desired outcome of the combat operation as it may relate to these categories as a desired endstate. Choosing a war termination category as part of the process in developing an objectives tree format as defined in the methodology presented in this chapter should guide the strategist to provide a feasible endstate to the operation. As a review, Clarke defines six major ways that disputes or conflicts can end, or in other words, war terminates:[8]

1. Armistice, truces, and cease-fires
2. Formal peace treaties
3. Joint published agreements
4. Declaration of a unilateral victory by the victor
5. Capitulation by the loser
6. Withdrawal of one of the parties.

Where he defines war termination, Clarke doesn't go beyond the cessation of hostilities. It is now clear that Clarke is merely defining goals for the military operations and not for the WoG concept required for ending a conflict or bringing it to a satisfactory result. As stated earlier, matching a conflicts war termination with a proposed "conflict resolution," the planner can choose from the following:

- Nation-building
- Reconstruction
- Economic development
- Humanitarian relief
- Transitioning security to the indigenous force
- Establishing a democratic nation
- Countering an insurgency and maintaining a lasting presence

- Invading a country to establish an imperial footprint or continual military presence
- Providing persistent foreign internal defense
- Withdrawing under fire without a deliberate transition plan.

The challenge remains, however, to know how to now deal with objectives that change in the middle of an ongoing operation and translating political goals into a workable endstate. As the objectives change, political and military leaders must constantly re-evaluate the endstate to align it with the new objectives.[9]

Epilogue

Si vis pacem, para bellum.
—Latin adage for "If you wish for peace, prepare for war"

BEYOND THE GUNS AND STEEL

Although critics will argue this is a scientific approach to bringing termination to a fight and a conflict to resolution, I emphasize that this is merely a tool or a set of tools to help strategists and planners as well as leaders to convey a methodical and visual way of developing a complete plan. My goal is to give the reader options to look at combat operations holistically and not just revert to the focus of gun fighting and closing with and destroying the enemy. It is clear that conflict is more than just ridding a foe of a capability; it has a postconflict requirement starting with war termination and leading into conflict resolution resulting in a desired endstate. In short, without proper planning our endstate will have an apogee of hope and a perigee of despair.

Although the Armed Forces have proven themselves a capable policy instrument, the nation has always struggled with conflict termination. The United States has often prevailed militarily while failing to achieve policy goals quickly and efficiently. A scan of joint publications suggests that military professionals embrace the idea of a termination strategy, but doctrine offers little practical help. It is time to take the next step: creating an interagency organization and practices that can effectively conduct termination planning.[1]

History, as I hope this work has shown, is replete with examples of the good, bad, and indifferent when it comes to viewing conflict as a life cycle process. Time progresses; the Middle East, although not at peace, is stable as Iraq is transitioned to a host-nation run security force, government, and society. The United States, the major fighting force in the effort to win the world's wars, has transformed to meet the conflicts of the future by integrating soft and hard power into its mindset when identifying resources. The next step, as defined in this text, is to integrate the interagency process alongside the military to ensure the required skills are available to plan and execute what comes after a fight.

Although doctrine and theory are guides, no formulas exist that will always lead to favorable conflict resolution. The art of planning military operations requires close coordination from a staff accustomed to working together all the time. The art of transition planning requires nothing less. If anything, transition strategies are more difficult because they must incorporate all instruments of national power in a coherent, synchronized fashion. The literature uses the terms *war termination* and *conflict termination* interchangeably. War is a specific type of violent conflict between two or more parties. Nations can have conflicts with each other that do not involve active, armed hostilities. Trade embargoes, retaliatory tariff barriers, the withdrawal of ambassadors, or sanction enforcement fall into this category. *Conflict resolution*, another term often used in the literature, is the group of processes that removes tensions between states or maintains them at levels consistent with the continued peaceful pursuit by states of their individual and collective goals.[2]

SOME GUIDELINES FOR CAMPAIGN PLANNERS[3]

The central argument throughout this text has been that the current gap in our operational doctrines regarding conflict termination seriously hampers our ability to plan effective military campaigns. Working from commonly accepted war termination precepts at the strategic level and armed with an appreciation of war termination issues in recent conflicts, let us propose some tentative first steps toward an appropriate doctrine in this arena.

- Identify a distinct war termination phase in the campaign planning process. Simply stated, war termination is too fundamental an issue to be subordinated as a lesser included component of some other aspect of the campaign planning process.
- Emphasize a regressive (i.e., backward-planning) approach to campaign development. Ever mindful of Iklé's caution that decision makers not take the first step toward war without considering the last, every aspect of a campaign plan—target selection, rules of engagement, psychological operations, to cite but a few examples—should be designed and evaluated according to contributions made or the effect on the explicitly defined endstate to be achieved. This can be accomplished most efficiently in a regressive planning sequence.
- Define the operational conditions to be produced during the terminal phase of the campaign in explicit, unambiguous terms. The absence of definition or detail in operational objectives may produce unintended consequences in the course

of a campaign. More important, the process of defining operational objectives with a high degree of clarity should prompt increased communication between the civilian and military leadership that will help to ensure congruence between operational objectives and the larger policy aims of a campaign.

- Consider establishing (in consultation with appropriate civilian decision makers) operational objectives that exceed the baseline political objectives of the campaign. Remember that the war termination process is a part of a larger implicit bargaining process, even while hostilities continue, and that the military contribution can significantly affect the leverage available to influence that process. This may include the seizure of territory or other high-value objectives whose possession would enhance our government's ability to secure a favorable political outcome.

- Consider how efforts to eliminate or degrade your opponent's command and control may affect, positively or negatively, your efforts to achieve particular objectives. Will your opponent be able to affect a cease-fire or otherwise control the actions of his forces? Efforts to target command and control facilities should carefully consider the trade-offs involved: At what point might the military advantages to be gained from targeting command and control be outweighed by potentially adverse effects on the goal of stopping the fighting?

- Consider the manner in which the tempo of the terminal phase of an operational campaign affects the ability to achieve established policy objectives. Once a campaign has reached the point of irreversibility, experience suggests that aggressive exploitation is most likely to secure the desired objectives at lower costs. A high operational tempo continues the pressure on the adversary and makes it more difficult for him to engage in destructive acts that raise the costs of your victory. Witness, for example, the wanton Iraqi destruction of Kuwaiti oil fields in the Gulf War. At other times, depending on one's knowledge of the enemy's decision process, it may be prudent to consider pauses or break-points in the terminal phases of a campaign to allow the opponent's decision-making machinery opportunities to cease fighting. This may be especially important where degraded command and control hampers an opponent's ability to consider and communicate a decision to quit. In either circumstance, operational tempo is clearly a key consideration in war termination.

- View war termination not as the end of hostilities but as the transition to a new postconflict phase characterized by both civil and military problems. This consideration implies an especially important role for various civil affairs functions. It also implies a requirement to plan the interagency transfer of certain responsibilities to national, international, or nongovernmental agencies. Effective battle hand-off requires planning and coordination before the fact.

Viewed independently, none of these proposed guidelines appears startling; some will even suggest, quite correctly, that these prescriptions are, like so many other aspects of warfighting doctrine, more exemplary of common sense than of any particular revealed wisdom. What is startling, however, is the absence of any fully developed approach to conflict termination in our current warfighting doctrines.

At this point it seems worth repeating our refrain that the war termination component of a campaign plan represents a transitional phase: a transition from war to peace, from a military-dominant role toward a civilian-dominant role, from a set of circumstances and problems generally familiar to operational planners toward others with which they may be decidedly less familiar. These points reinforce the importance of a high level of dialogue and coordination between civilian and military decision makers regarding the conflict termination process. As Fred Iklé notes, "In preparing a major military operation, military leaders and civilian officials can effectively work together . . . to create a well-meshed, integrated plan."[4] The ability of military leaders to contribute to that joint planning process will in part depend on the extent to which they have carefully considered the challenges posed by the war termination problem in the period before deterrence fails.

Ambassador Ryan Crocker ended his distinguished stint in Iraq with these fitting words: "In the end, how we leave and what we leave behind will be more important than how we came."[5] Translating the good Ambassador's comment: War Termination + Conflict Resolution = a feasible and desired endstate to any fight.

Agreement on Ending the War and Restoring Peace in Viet-Nam

The Parties participating in the Paris Conference on Viet-Nam,

With a view to ending the war and restoring peace in Viet-Nam on the basis of respect for the Vietnamese people's fundamental national rights and the South Vietnamese people's right to self-determination, and to contributing to the consolidation of peace in Asia and the world,

Have agreed on the following provisions and undertake to respect and to implement them:

CHAPTER I

The Vietnamese People's Fundamental National Rights

Article 1

The United States and all other countries respect the independence, sovereignty, unity, and territorial integrity of Viet-Nam as recognized by the 1954 Geneva Agreements on Viet-Nam.

CHAPTER II

Cessation Of Hostilities—Withdrawal of Troops

Article 2

A cease-fire shall be observed throughout South Viet-Nam as of 2400 hours G.M.T. [Greenwich Mean Time], on January 27, 1973.

At the same hour, the United States will stop all its military activities against the territory of the Democratic Republic of Viet-Nam by ground, air and naval forces, wherever they may be based, and end the mining of the territorial waters, ports, harbors, and waterways of the Democratic Republic of Viet-Nam. The United States will remove, permanently deactivate or destroy all the mines in the territorial waters, ports, harbors, and waterways of North Viet-Nam as soon as this Agreement goes into effect.

The complete cessation of hostilities mentioned in this Article shall be durable and without limit of time.

Article 3

The parties undertake to maintain the cease-fire and to ensure a lasting and stable peace.

As soon as the cease-fire goes into effect: (a) The United States forces and those of the other foreign countries allied with the United States and the Republic of Viet-Nam shall remain in-place pending the implementation of the plan of troop withdrawal. The Four-Party Joint Military Commission described in Article 16 shall determine the modalities.

(b) The armed forces of the two South Vietnamese parties shall remain in-place. The Two-Party Joint Military Commission described in Article 17 shall determine the areas controlled by each party and the modalities of stationing.

(c) The regular forces of all services and arms and the irregular forces of the parties in South Viet-Nam shall stop all offensive activities against each other and shall strictly abide by the following stipulations:

– All acts of force on the ground, in the air, and on the sea shall be prohibited;

– All hostile acts, terrorism and reprisals by both sides will be banned.

Article 4

The United States will not continue its military involvement or intervene in the internal affairs of South Viet-Nam.

Article 5

Within sixty days of the signing of this Agreement, there will be a total withdrawal from South Viet-Nam of troops, military advisers, and military personnel, including technical military personnel and military personnel associated with the pacification program, armaments, munitions, and war material of the United States and those of the other foreign countries mentioned in Article 3 (a). Advisers from the above-mentioned countries to all paramilitary organizations and the police force will also be withdrawn within the same period of time.

Article 6

The dismantlement of all military bases in South Viet-Nam of the United States and of the other foreign countries mentioned in Article 3 (a) shall be completed within sixty days of the signing of this agreement.

Article 7

From the enforcement of the cease-fire to the formation of the government provided for in Article 9 (b) and 14 of this Agreement, the two South Vietnamese parties shall not accept the introduction of troops, military advisers, and military personnel including technical military personnel, armaments, munitions, and war material into South Viet-Nam.

The two South Vietnamese parties shall be permitted to make periodic replacement of armaments, munitions and war material which have been destroyed, damaged, worn out or used up after the cease-fire, on the basis of piece-for-piece, of the same characteristics and properties, under the supervision of the Joint Military Commission of the two South Vietnamese parties and of the International Commission of Control and Supervision.

The Return of Captured Military Personnel and Foreign Civilians and Captured and Detained Vietnamese Civilian Personnel

Article 8

(a) The return of captured military personnel and foreign civilians of the parties shall be carried out simultaneously with and completed not later than the same day as the troop withdrawal mentioned in Article 5. The parties shall exchange complete lists of the above-mentioned captured military personnel and foreign civilians on the day of the signing of this Agreement.

(b) The parties shall help each other to get information about those military personnel and foreign civilians of the parties missing in action, to determine the location and take care of the graves of the dead so as to facilitate the exhumation and repatriation of the remains, and to take any such other measures as may be required to get information about those still considered missing in action.

(c) The question of the return of Vietnamese civilian personnel captured and detained in South Viet-Nam will be resolved by the two South Vietnamese parties on the basis of the principles of Article 21 (b) of the Agreement on the Cessation of Hostilities in Viet-Nam of July 20, 1954. The two South Vietnamese parties will do so in a spirit of national reconciliation and concord, with a view to ending hatred and enmity, in order to ease suffering and to reunite families. The two South Vietnamese parties will do their utmost to resolve this question within ninety days after the cease-fire comes into effect.

CHAPTER IV

The Exercise of the South Vietnamese People's Right to Self-Determination

Article 9

The Government of the United States of America and the Government of the Democratic Republic of Viet-Nam undertake to respect the following principles for the exercise of the South Vietnamese people's right to self-determination:

(a) The South Vietnamese people's right to self-determination is sacred, inalienable, and shall be respected by all countries.

(b) The South Vietnamese people shall decide themselves the political future of South Viet-Nam through genuinely free and democratic general elections under international supervision.

(c) Foreign countries shall not impose any political tendency or personality on the South Vietnamese people.

Article 10

The two South Vietnamese parties undertake to respect the cease-fire and maintain peace in South Viet-Nam, settle all matters of contention through negotiations, and avoid all armed conflict.

Article 11

Immediately after the cease-fire, the two South Vietnamese parties will:

– achieve national reconciliation and concord, end hatred and enmity, prohibit all acts of reprisal and discrimination against individuals or organizations that have collaborated with one side or the other;

– ensure the democratic liberties of the people: personal freedom, freedom of speech, freedom of the press, freedom of meeting, freedom of organization, freedom of political activities, freedom of belief, freedom of movement, freedom of residence, freedom of work, right to property ownership, and right to free enterprise.

Article 12

(a) Immediately after the cease-fire, the two South Vietnamese parties shall hold consultations in a spirit of national reconciliation and concord, mutual respect, and mutual non-elimination to set up a National Council of National Reconciliation and Concord of three equal segments. The Council shall operate on the principle of unanimity, After the National Council of National Reconciliation and Concord has assumed its functions, the two South Vietnamese parties will consult about the formation of councils at lower levels. The two South Vietnamese parties shall sign an agreement on the internal matters of South Viet-Nam as soon as possible and do their utmost to accomplish this within ninety days after the cease-fire comes into effect, in keeping with the South Vietnamese people's aspirations for peace, independence and democracy.

(b) The National Council of National Reconciliation and Concord shall have the task of promoting the two South Vietnamese parties' implementation of this Agreement, achievement of national reconciliation and concord and ensurance of democratic liberties. The National Council of National Reconciliation and Concord will organize the free and democratic general elections provided for in Article 9 (b) and decide the procedures and modalities of these general elections. The institutions for which the general elections are to be held will be agreed upon through consultations between the two South Vietnamese parties. The National Council of National Reconciliation and Concord will also decide the procedures and modalities of such local elections as the two South Vietnamese parties agree upon.

Article 13

The question of Vietnamese armed forces in South Viet-Nam shall be settled by the two South Vietnamese parties in a spirit of national reconciliation and concord, equality and mutual respect, without foreign interference, in accordance with the postwar situation. Among the questions to be discussed by the two South Vietnamese parties are steps to reduce their military effectives and to demobilize the troops being reduced. The two South Vietnamese parties will accomplish this as soon as possible.

Article 14

South Viet-Nam will pursue a foreign policy of peace and independence. It will be prepared to establish relations with all countries irrespective of their political and social systems on the basis of mutual respect for independence and sovereignty and accept economic and technical aid from any country with no political conditions attached. The acceptance of military aid by South Viet-Nam in the future shall come under the authority of the government set up after the general elections in South Viet-Nam provided for in Article 9 (b).

CHAPTER V

The Reunification of Viet-Nam and the Relationship Between North and South Viet-Nam

Article 15

The reunification of Viet-Nam shall be carried out step by step through peaceful means on the basis of discussions and agreements between North and South Viet-Nam, without coercion or annexation by either party, and without foreign interference. The time for reunification will be agreed upon by North and South Viet-Nam. Pending reunification:

(a) The military demarcation line between the two zones at the 17th parallel is only provisional and not a political or territorial boundary, as provided for in paragraph 6 of the Final Declaration of the 1954 Geneva Conference.

(b) North and South Viet-Nam shall respect the Demilitarized Zone on either side of the Provisional Military Demarcation Line.

(c) North and South Viet-Nam shall promptly start negotiations with a view to reestablishing normal relations in various fields. Among the questions to be negotiated are the modalities of civilian movement across the Provisional Military Demarcation Line,

(d) North and South Viet-Nam shall not join any military alliance or military bloc and shall not allow foreign powers to maintain military bases, troops; military advisers, and military personnel on their respective territories, as stipulated in the 1954 Geneva Agreements on Viet-Nam.

The Joint Military Commissions, The International Commission of Control and Supervision, The International Conference

Article 16

(a) The Parties participating in the Paris Conference on Viet-Nam shall immediately designate representatives to form a Four-Party Joint Military Commission with the task of ensuring joint action by the parties in implementing the following provisions of this Agreement:

  – The first paragraph of Article 2, regarding the enforcement of the cease-fire throughout South Viet-Nam;

- Article 3 (a), regarding the cease-fire by U.S. forces and those of the other foreign countries referred to in that Article;
- Article 3 (c), regarding the cease-fire between all parties in South Viet-Nam;
- Article 5, regarding the withdrawal from South Viet-Nam of U.S. troops and those of the other foreign countries mentioned in Article 3 (a);
- Article 6, regarding the dismantlement of military bases in South Viet-Nam of the United States and those of the other foreign countries mentioned in Article 3 (a);
- Article 8 (a), regarding the return of captured military personnel and foreign civilians of the parties;
- Article 8 (b), regarding the mutual assistance of the parties in getting information about those military personnel and foreign civilians of the parties missing in action.

(b) The Four-Party Joint Military Commission shall operate in accordance with the principle of consultations and unanimity. Disagreements shall be referred to the International Commission of Control and Supervision.

(c) The Four-Party Joint Military Commission shall begin operating immediately after the signing of this Agreement and end its activities in sixty days, after the completion of the withdrawal of U.S. troops and those of the other foreign countries mentioned in Article 3 (a) and the completion of the return of captured military personnel and foreign civilians of the parties.

(d) The four parties shall agree immediately on the organization, the working procedure, means of activity, and expenditures of the Four-Party Joint Military Commission.

Article 17

(a) The two South Vietnamese parties shall immediately designate representatives to form a Two-Party Joint Military Commission with the task of ensuring joint action by the two South Vietnamese parties in implementing the following provisions of this Agreement:

- The first paragraph of Article 2, regarding the enforcement of the cease-fire throughout South Viet-Nam, when the Four-Party Joint Military Commission has ended its activities;
- Article 3 (b), regarding the cease-fire between the two South Vietnamese parties;
- Article 3 (c), regarding the cease-fire between all parties in South Viet-Nam, when the Four-Party Joint Military Commission has ended its activities;
- Article 7, regarding the prohibition of the introduction of troops into South Viet-Nam and all other provisions of this Article;
- Article 8 (c), regarding the question of the return of Vietnamese civilian personnel captured and detained in South Viet-Nam;

- Article 13, regarding the reduction of the military effectives of the two South Vietnamese parties and the demobilization of the troops being reduced.

(b) Disagreements shall be referred to the International Commission of Control and Supervision.

(c) After the signing of this Agreement, the Two-Party Joint Military Commission shall agree immediately on the measures and organization aimed at enforcing the cease-fire and preserving peace in South Viet-Nam,

Article 18

(a) After the signing of this Agreement, an International Commission of Control and Supervision shall be established immediately.

(b) Until the International Conference provided for in Article 19 makes definitive arrangements, the International Commission of Control and Supervision will report to the four parties on matters concerning the control and supervision of the implementation of the following provisions of this Agreement:

- The first paragraph of Article 2, regarding the enforcement of the cease-fire throughout South Viet-Nam;
- Article 3 (a), regarding the cease-fire by U.S. forces and those of the other foreign countries referred to in that Article;
- Article 3 (c), regarding the cease-fire between all the parties in South Viet-Nam;
- Article 5, regarding the withdrawal from South Viet-Nam of U.S. troops and those of the other foreign countries mentioned in Article 3 (a);
- Article 6, regarding the dismantlement of military bases in South Viet-Nam of the United States and those of the other foreign countries mentioned in Article 3 (a);
- Article 8 (a), regarding the return of captured military personnel and foreign civilians of the parties.

The International Commission of Control and Supervision shall form control teams for carrying out its tasks. The four parties shall agree immediately on the location and operation of these teams. The parties will facilitate their operation.

(c) Until the International Conference makes definitive arrangements, the International Commission of Control and Supervision will report to the two South Vietnamese parties on matters concerning the control and supervision of the implementation of the following provisions of this Agreement:

- The first paragraph of Article 2, regarding the enforcement of the cease-fire throughout South Viet-Nam, when the Four-Party Joint Military Commission has ended its activities;
- Article 3 (b), regarding the cease-fire between the two South Vietnamese parties;
- Article 3 (c), regarding the cease-fire between all parties in South Viet-Nam, when the Four-Party Joint Military Commission has ended its activities;

- Article 7, regarding the prohibition of the introduction of troops into South Viet-Nam and all other provisions of this Article;
- Article 8 (c), regarding the question of the return of Vietnamese civilian personnel captured and detained in South Viet-Nam;
- Article 9 (b), regarding the free and democratic general elections in South Viet-Nam;
- Article 13, regarding the reduction of the military effectives of the two South Vietnamese parties and the demobilization of the troops being reduced.

The International Commission of Control and Supervision shall form control teams for carrying out its tasks. The two South Vietnamese parties shall agree immediately on the location and operation of these teams. The two South Vietnamese parties will facilitate their operation.

(d) The International Commission of Control and Supervision shall be composed of representatives of four countries: Canada, Hungary, Indonesia and Poland. The chairmanship of this Commission will rotate among the members for specific periods to be determined by the Commission.

(e) The International Commission of Control and Supervision shall carry out its tasks in accordance with the principle of respect for the sovereignty of South Viet-Nam.

(f) The International Commission of Control and Supervision shall operate in accordance with the principle of consultations and unanimity.

(g) The International Commission of Control and Supervision shall begin operating when a cease-fire comes into force in Viet-Nam. As regards the provisions in Article 18 (b) concerning the four parties, the International Commission of Control and Supervision shall end its activities when the Commission's tasks of control and supervision regarding these provisions have been fulfilled. As regards the provisions in Article 18 (c) concerning the two South Vietnamese parties, the International Commission of Control and Supervision shall end its activities on the request of the government formed after the general elections in South Viet-Nam provided for in Article 9 (b).

(h) The four parties shall agree immediately on the organization, means of activity, and expenditures of the International Commission of Control and Supervision. The relationship between the International Commission and the International Conference will be agreed upon by the International Commission and the International Conference.

Article 19

The parties agree on the convening of an International Conference within thirty days of the signing of this Agreement to acknowledge the signed agreements; to guarantee the ending of the war, the maintenance of peace in Viet-Nam, the respect of the Vietnamese people's fundamental national rights, and the South Vietnamese people's right to self-determination; and to contribute to and guarantee peace in Indochina.

The United States and the Democratic Republic of Viet-Nam, on behalf of the parties participating in the Paris Conference on Viet-Nam will propose to the following parties that they participate in this International Conference: the People's Republic of China, the Republic of France, the Union of Soviet Socialist Republics, the United Kingdom, the four countries of the International Commission of Control and Supervision, and the Secretary General of the United Nations, together with the parties participating in the Paris Conference on Viet-Nam.

CHAPTER VII

Regarding Cambodia and Laos

Article 20

(a) The parties participating in the Paris Conference on Viet-Nam shall strictly respect the 1954 Geneva Agreements on Cambodia and the 1954 Geneva Agreements on Laos, which recognized the Cambodian and the Lao peoples' fundamental national rights, i.e., the independence, sovereignty, unity, and territorial integrity of these countries. The parties shall respect the neutrality of Cambodia and Laos.

The parties participating in the Paris Conference on Viet-Nam undertake to refrain from using the territory of Cambodia and the territory of Laos to encroach on the sovereignty and security of one another and of other countries.

(b) Foreign countries shall put an end to all military activities in Cambodia and Laos, totally withdraw from and refrain from reintroducing into these two countries troops, military advisers and military personnel, armaments, munitions and war material.

(c) The internal affairs of Cambodia and Laos shall be settled by the people of each of these countries without foreign interference.

(d) The problems existing between the Indochinese countries shall be settled by the Indochinese parties on the basis of respect for each other's independence, sovereignty, and territorial integrity, and non-interference in each other's internal affairs.

CHAPTER VIII

The Relationship Between the United States and the Democratic Republic of Viet-Nam

Article 21

The United States anticipates that this Agreement will usher in an era of reconciliation with the Democratic Republic of Viet-Nam as with all the peoples of Indochina. In pursuance of its traditional policy, the United States will contribute to healing the wounds of war and to postwar reconstruction of the Democratic Republic of Viet-Nam and throughout Indochina.

Article 22

The ending of the war, the restoration of peace in Viet-Nam, and the strict implementation of this Agreement will create conditions for establishing a new, equal and mutually beneficial relationship between the United States and the Democratic Republic of Viet-Nam on the basis of respect for each other's independence and sovereignty, and non-interference in each other's internal affairs. At the same time this will ensure stable peace in Viet-Nam and contribute to the preservation of lasting peace in Indochina and Southeast Asia.

CHAPTER IX

Other Provisions

Article 23

This Agreement shall enter into force upon signature by plenipotentiary representatives of the parties participating in the Paris Conference on Viet-Nam. All the parties concerned shall strictly implement this Agreement and its Protocols. Done in Paris this twenty-seventh day of January, one thousand nine hundred and seventy-three, in English and Vietnamese. The English and Vietnamese texts are official and equally authentic.

For the Government of the for the Government of the United States of America: Republic of Viet-Nam:

(Signed):   *(Signed):*

William P. Rogers   Tran Van Lam
Secretary of State   Minister for Foreign Affairs

For the Government of the for the Provisional Democratic Republic Revolutionary Government of Viet-Nam: of the Republic of South Viet-Nam:

(Signed):   *(Signed):*

Nguyen Duy Trinh   Nguyen Thi Binh
Minister for Foreign Affairs Minister for Foreign Affairs

Notes

INTRODUCTION

1. Description offered by then-Lieutenant Margaret "Meg" Fitzpatrick stationed in South Baghdad from September 2007 until November 2008 in what was formerly referred to as "The Triangle of Death." Interview in September 2009 at Camp Striker, Baghdad, Iraq.

2. Defense Secretary Robert M. Gates, Landon Lecture (Kansas State University), Remarks as Delivered by Secretary of Defense, Manhattan, Kansas, Monday, November 26, 2007 at http://www.defense.gov/speeches/speech.aspx?speechid=1199.

3. Kimberly Kagan, "The Patton of Counterinsurgency." Institute for the Study of War home page at www.understandingwar.org/other-work/patton-counterinsurgency (accessed January 8, 2010).

4. Carl von Clausewitz, *On War,* ed. and trans. Michael Howard and Peter Paret (Princeton, NJ: Princeton University Press, 1976), 80.

5. For this study I will use the terms *war termination* and *conflict resolution* as dependent terms. *Endstate* may be used synonymously as a goal in terminating conflict. The reader may also see the word *dispute* used interchangeably for *war, foray*, and *conflict*.

6. As the fight in Iraq transitions to the civil authorities, General Raymond T. Odierno often tells congressional leaders and other influential people that his biggest fear is that the resource providers will fall short in supporting the war's goals as it transitions from a military to a civil lead. The example he openly uses is that of the 2007 film *Charlie Wilson's War* where the Congressman Charlie Wilson, a Democrat from Texas (played by actor Tom Hanks) has supported the resourcing for post-Soviet occupation of Afghanistan, but finds almost no enthusiasm in the U.S. government for even the modest measures he proposes as the war comes to an "end," and he asks for a few million dollars to support the building of schools

only be told that the war is over. The war terminates, but the conflict resolution requirements are ignored.

7. George Friedman, *The Next 100 Years: A Forecast for the 21st Century* (New York: Anchor Books, 2010), 101.

8. *Balkanization* is defined as a geographical term originally used to describe the fragmentation of the Balkans at the end of the 20th century conflict. In one sense, it is the fragmentation of regional trisections into Sunni, Shi'a, and Kurdish political entities. For the purpose of this work, however, *Balkanization* describes a by-product of how military units were place in a timeline against a regional area during the conflict first in the Balkans and then later in Iraq. To be more specific, an example is when units from the U.S. military are scheduled to occupy a set forward operating environment and its accompanying bases and other position in the U.S. Division regional areas. The one-for-one swap in mission and location is what is often referred to as *Balkanization*.

9. http://www.newyorker.com/talk/2009/01/26/090126ta_talk_hertzberg.

10. Brig. Gen. David L. Grange, and Scott Swanson, "Confronting Irregular Challenges in the 21st Century," Irregular Warfare Concept Series: Whole of World Collaboration (March 10, 2009), 1.

11. Michael E. Brown, J. Coté, R. Owen, Sean M. Lynn-Jones., and Steven E Miller, eds., New Global Dangers: Changing Dimensions of International Security (Cambridge, MA: MIT Press, 2004), 39.

12. Bruce B.G. Clarke, *Conflict Termination: A Rational Model* (Carlisle Barracks, PA: Strategic Studies Institute, U.S. Army War College, 1992), 9.

13. Grange and Swanson, "Confronting Irregular Challenges," 3.

14. *Mission creep* is the expansion of a project or mission beyond its original goals, often after initial successes. The term often implies a certain disapproval of newly adopted goals by the user of the term. Mission creep is usually considered undesirable due to the dangerous path of each success breeding more ambitious attempts, only stopping when a final, often catastrophic, failure occurs. The term was originally applied exclusively to military operations, but has recently been applied to many different fields, mainly the growth of bureaucracies. Although this term got its popularity in use in the post-Somalia debacle of October 1993, the concept has obviously been around since the beginning of time. I equate mission creep to a poor sense of proper planning where consequence management (another popular term used to ensure that mistakes are rectified by managing their outcome) is thought of a deliberate step in the process as opposed to meaning a poor job in planning properly for the consequences of a mission set. James D. Hessman, "Three Decades of Mission Creep; Loy: The 'Do More With Less' Well Has Run Dry," http://www.navyleague.org/seapower/three_decades_of _mission_creep.htm.

15. The Coalition formed in 1990 to thwart Saddam Hussein's armed invasion of Kuwait consisted of 32 coalition partners.

16. Frederick J. Ourso, *War Termination: Do Planning Principles Change with the Nature of the War?* (Newport, RI: Naval War College, May 18, 1998), 6–7.

17. The idea that nationalist struggles can overthrow imperialism shows a failure to understand what imperialism is. The basis of imperialism is the division of the world into an "anarchic" system of independent nation-states. There is no

larger structure of decision making that regulates human society on a global scale. Nation-states are thus only constrained in their conduct on the world stage by fear of what other states can do to them. Competition between nation-states puts pressure on each state to maximize its power to avoid subordination to others. States that have little power will be under severe pressure to align themselves with more muscular states that have major military and economic forces at their disposal. Blog entry at http://www.uncanny.net/~wsa/iraq3.html.

18. Fred Charles Iklé, *Every War Must End* (New York: Columbia University Press, 1971), 519; citing General Colin Powell in his autobiography, *My American Journey*, 106.

19. Ibid., 2.

20. Clarke, "Conflict Termination," 3.

21. Iklé, *Every War Must End*, 7.

22. Clarke, "Conflict Termination," 9–10 and William O. Staudenmaier, "Conflict Termination in the Nuclear Era," in *Conflict Termination and Military Strategy: Coercion, Persuasion and War*, ed. Stephen J. Cimbala and Keith A. Dunn (Boulder, CO: Westview Press, 1987), 233–236.

23. Iklé, *Every War Must End*, ix.

24. Bruce C. Bade, *War Termination: Why Don't We Plan for It?* (Carlisle Barracks, PA: U.S. National War College, 1994), 1.

25. Ibid., 16.

26. Ibid., 15.

27. Clarke, "Conflict Termination," 2.

28. "Active Component Stress on the Force" (Alexandria, VA: U.S. Army, Deputy Chief of Staff for Operations, Training, and Analysis Division, July 2009).

29. The military calls resourcing (money, time, equipment, people, information) a "means" to achieve an end.

30. Gates, Landon Lecture.

31. Grange and Swanson, "Confronting Irregular Challenges," 1.

32. Gates, Landon Lecture, 6.

33. Ibid., 8.

34. Friedman, *The Next 100 Years*, 18, 22.

CHAPTER 1

1. Jim Heintz, "On War's Anniversary, Georgia, Russia Vie in Media: Georgia, Russia Push for Public Opinion Victory on War's Anniversary," The Associated Press for ABC International, August 6, 2009, http://abcnews.go.com/International/wireStory?id=8266942.

2. Robert A. Coalson, "Year after Russia-Georgia War—A New Reality, but Old Relations," *Radio Free Europe, Radio Liberty*, April 29, 2010, http://www.rferl.org/content/A_Year_After_RussiaGeorgia_War__A_New_Reality_But_Old_Relations/1793048.html.

3. Carl von Clausewitz, *On War*, ed. and trans. Michael Howard and Peter Paret. Princeton, NJ: Princeton University Press, 1984, Book 2, Chapter 2, Paragraph 24.

4. William R. Gruver, "Fog of War: Why Clausewitz Would Not Be Happy with Obama's New Afghanistan Strategy." *The New Republic* (December 8, 2009), http://www.tnr.com/article/world/fog-war.

5. Alistair Morley, *Historical Analysis of Conflict Termination* (Fornborough, Hants, UK: Policy and Capabilities Studies Department, Defense Science and Technology Laboratory), http://www.ima.org.uk/conflict/papers/Morley.pdf., p. 1.

6. Ibid.

7. Ibid., 2.

8. Ibid.

9. A useful objection here is that the scheme of utility for the policy elite may not be synonymous with that of the state. Morley, "Historical Analysis," 2.

10. Ibid.

11. Ibid.

12. An attempt to ascertain the mental states of the policy elites involved from historical records would be fraught with potential error. However, the empirical condition of their states, and the objective held, are much less in dispute among historians. Hence, the use of existing costs and benefits at war termination are preferred as indicating the direction in which further fighting would have progressed. Ibid.

13. Fred Charles Iklé, *Every War Must End* (New York: Columbia University Press, 1971), ix.

14. Colin L. Powell, *My American Journey* (New York: Random House, 1995), 519.

15. Operation Desert Storm: Evaluation of the Air Campaign (Letter Report, 06/12/97, GAO/NSIAD-97-134); Appendix V; http://www.fas.org/man/gao/nsiad97134/app_05.htm.

16. The Shi'a uprising following the Coalition ground attack in spring 1991 was a sign of things to come with Operation Iraqi Freedom and the insurgent efforts by Shi'a elements and the vulnerability of President Saddam Hussein's regime.

17. Harry R. Yager, *Strategic Theory for the 21st Century: The Little Book on Big Strategy* (Carlisle, PA: Strategic Studies Institute, February 2006), 6.

18. Ibid.

19. Bruce B.G. Clarke, *Conflict Termination: A Rational Model* (Carlisle Barracks, PA, U.S. Army War College, 1992), 9-10; William O. Staudenmaier, "Conflict Termination in the Nuclear Era," in *Conflict Termination and Military Strategy: Coercion, Persuasion and War*, ed. Stephen J. Cimbala and Keith A. Dunn, 233–236 (Boulder, CO: Westview Press, 1987).

20. Patrick Cockburn, "Sadr Calls Six-Month Ceasefire to Prevent Civil War," *The Independent World* (August 30, 2007), http://www.independent.co.uk/news/world/middle-east/sadr-calls-sixmonth-ceasefire-to-prevent-civil-war-463540.html.

21. "Text of the Korean War Armistice Agreement," July 27, 1953, http://news.findlaw.com/hdocs/docs/korea/kwarmagr072753.html (accessed December 24, 2009).

22. "The Armistice," *The War to End All Wars*, www.FirstWorldWar.com. http://www.firstworldwar.com/features/armistice.htm (accessed December 27, 2009).

23. "What Is Remembrance Day?" CBBC Newsround. http://news
.bbc.co.uk/cbbcnews/hi/find_out/guides/uk/remembrance_day/
newsid_2438000/2438201.stm (accessed December 27, 2009).

24. "1949 Armistice," *Middle East, Land of Conflict.* CNN. http://www.cnn
.com/SPECIALS/2001/mideast/stories/history.maps/armistice.html   (accessed
December 27, 2009).

25. Racueils de la Societe Internationale de Droit Penal Militaire et de Droit
de la Guerre, Vol. II: Deuxieme Congres International, Florence, 17–20 May 1961;
and Oliver Thatcher Jr, and Edgar Holmes McNeal, eds., *A Source Book for Medieval
History* (New York: Scribners, 1905), 417–418, http://www.fordham.edu/halsall/
source/t-of-god.html. Subtitled, *Medieval Sourcebook: Truce of God—Bishopric of
Terouanne,* 1063. In this document it states that Drogo, bishop of Terouanne, and
count Baldwin [of Hainault] have established this peace with the cooperation of
the clergy and people of the land.

Dearest brothers in the Lord, these are the conditions which you must
observe during the time of the peace which is commonly called the truce of
God, and which begins with sunset on Wednesday and lasts until sunrise
on Monday.

1. During those four days and five nights no man or woman shall assault,
wound, or slay another, or attack, seize, or destroy a castle, burg, or villa, by
craft or by violence.

2. If anyone violates this peace and disobeys these commands of ours, he
shall be exiled for thirty years as a penance, and before he leaves the bishopric he shall make compensation for the injury which he committed. Otherwise he shall be excommunicated by the Lord God and excluded from all
Christian fellowship.

3. All who associate with him in any way, who give him advice or aid, or
hold converse with him, unless it be to advise him to do penance and to
leave the bishopric, shall be under excommunication until they have made
satisfaction.

4. If any violator of the peace shall fall sick and die before he completes his
penance, no Christian shall visit him or move his body from the place where
it lay, or receive any of his possessions.

5. In addition, brethren, you should observe the peace in regard to lands
and animals and all things that can be possessed. If anyone takes from
another an animal, a coin, or a garment, during the days of the truce, he
shall be excommunicated unless he makes satisfaction. If he desires to make
satisfaction for his crime he shall first restore the thing which he stole or its
value in money, and shall do penance for seven years within the bishopric.
If he should die before he makes satisfaction and completes his penance, his
body shall not be buried or removed from the place where it lay, unless his
family shall make satisfaction for him to the person whom he injured.

6. During the days of the peace, no one shall make a hostile expedition on
horseback, except when summoned by the count; and all who go with the
count shall take for their support only as much as is necessary for themselves and their horses.

7. All merchants and other men who pass through your territory from other lands shall have peace from you.

8. You shall also keep this peace every day of the week from the beginning of Advent to the octave of Epiphany and from the beginning of Lent to the octave of Easter, and from the feast of Rogations [the Monday before Ascension Day] to the octave of Pentecost.

9. We command all priests on feast days and Sundays to pray for all who keep the peace, and to curse all who violate it or support its violators.

10. If anyone has been accused of violating the peace and denies the charge, he shall take the communion and undergo the ordeal of hot iron. If he is found guilty, he shall do penance within the bishopric for seven years.

26. Hugo Grotius (1583-1645) [Hugo, Huigh or Hugeianus de Groot] was a towering figure in philosophy, law, political theory and associated fields during the 17th century and for hundreds of years afterward. His work ranged over a wide array of topics, though he is best known to philosophers today for his contributions to the natural law theories of normativity that emerged in the later medieval and early modern periods.

27. Sydney D. Bailey, "Cease-Fires, Truces, and Armistices in the Practices of the UN Security Council," *American Journal of International Law* 71 (1971), 461.

28. Gideon Boas, James L. Bischoff, and Natalie L. Reid, *Elements of Crimes under International Law*. International Criminal Law Practitioner Library. Vol. 2 (New York: Cambridge University Press, February 16, 2009), 2224.

29. Lieutenant General Sir Thomas Picton (August 1758–June 18, 1815) was a British Army officer from Wales who fought in a number of campaigns for Britain, and rose to the rank of lieutenant general. According to the historian Alessandro Barbero, Picton was "respected for his courage and feared for his irascible temperament." He is chiefly remembered for his exploits under the Duke of Wellington in the Iberian Peninsular War and at the Battle of Waterloo, where he was mortally wounded while his division stopped d'Erlon's corps attack against the allied center left, and so became the most senior officer to die at Waterloo.

30. Ultimo: in or of the month preceding the present one.

31. Santarém was involved in some of the most important events in Portuguese history in the 19th century. It was one of the main stages for the Peninsula wars and was besieged by the Duke of Wellington in 1810 to 1811, the event being related in 'Santarém or Sketches of manners and customs in the interior of Portugal," the narrative of the travels of a Scottish doctor and British army officer, John Gordon Smith. The town was also in the forefront of the Liberal struggles. Sá da Bandeira (Statesman and Military), Passos Manuel (Statesman and Military), and Braamcamp Freire (Politician) being leading liberals that were either born in or linked to Santarém. http://www.ribatejo.com/ecos/santarem/ingles/ihistori.html.

32. *The Dispatches of Field Marshall the Duke of Wellington during His Various Campaigns in India, Denmark, Portugal, Spain, the Low Countries, and France from 1799–1818*, Vols. 1 and 2 (Whitefish, MT: Kessinger Publishing, 2008), 642, http://www.zum.de/whkmla/period/17891914/x17891914.html.

33. Fort Pickens stood at the western edge of an island, running roughly parallel to the coastline and separated from it by Pensacola Bay. On January 10, 1861, the same day as Florida seceded from the Union, the small federal

contingent at Pensacola took steps to defend federal property. Lieutenant Adam J. Slemmer, in charge of army troops and acting under authority from Washington, transferred his command from the mainland to Fort Pickens, a more defensible position, which provided relative easy reinforcement from the Gulf of Mexico. Two days later, Florida and Alabama troops took over all the mainland posts, but failed to dislodge the federal presence at Fort Pickens. Toward the latter part of January, reinforcements commanded by Captain Israel Vogdes were sent to the fort aboard the USS *Brooklyn*, a powerful steam-powered warship. Additional naval support was also sent to Pensacola, including the recently built sailing frigate, the USS *Sabine*. Although these vessels arrived safely, the Brooklyn landed only provisions, not troops, at the fort. The explanation for this change of policy was an arrangement, or "truce," entered into by President Buchanan and Florida officials, by which Florida agreed not to attack the fort and, in return, the *Brooklyn* would not land its troops unless the fort were attacked or preparations made for its attack. Thus, an uneasy standoff existed at Fort Pickens, as the South put increasing military pressure on the fort, while a considerable Union military presence remained close by.

Meanwhile, at Charleston Harbor, a similar situation existed at Fort Sumter. Like Fort Pickens, Sumter was located offshore, being constructed on an artificial island made from the granite of northern quarries. Nearby fortifications, such as Forts Moultrie and Johnson, and Castle Pinckney, virtually surrounded it. Prior to South Carolina's secession on December 20, 1860, the Buchanan administration declined to reinforce the small federal contingent largely housed at Fort Moultrie, and ordered its commander, Major Robert Anderson, to defend the forts if attacked but not to provoke hostilities. After December 20, Anderson's situation became more difficult. With public sentiment pressing for action, South Carolina sent commissioners to Washington to negotiate the transfer of the forts to the state, and requested immediate control of Fort Sumter. Like Slemmer, Anderson considered his situation increasingly precarious, indeed untenable, if South Carolina occupied Sumter. After nightfall, on the evening of December 26, Anderson moved his small force from Moultrie to the more defensible Sumter.

Despite South Carolina's insistence that Anderson's action was a hostile act and must be repudiated, President Buchanan refused to order Anderson to return. South Carolina then proceeded to occupy federal property in Charleston, including the military posts surrounding Sumter. By January 1, only Sumter remained a Union outpost in the midst of secessionist South Carolina. Stiffening his resolve to protect Anderson's vulnerable garrison, President Buchanan approved an expedition headed by a chartered merchant steamer, the *Star of the West*, to resupply and reinforce Fort Sumter. On January 9, 1861, the ship arrived at Charleston Harbor, but turned back when it was fired on by South Carolina's batteries. Despite the outbreak of fighting, war did not ensue. As at Pensacola, a precarious truce went into effect in Charleston Harbor. The Confederate government, which assumed responsibility for Sumter after its establishment, tightened the noose around the fort, while the Union garrison continued to hold firm. The situation at Sumter received considerably more public attention, both in the North and the South, than that at Pickens. It rapidly became a symbol of rival definitions of sovereignty and honor. Tulane University's home page, "Crisis at Fort Sumter," section, http://www .tulane.edu/~sumter/Background/BackgroundForts.html.

34. David J. Coles, David Stephen Heidler, Jeanne T. Heidler, and James M. McPherson, *Encyclopedia of the American Civil War: A Political, Social, and Military History* (New York: W.W. Norton, 2002), 744.

35. Blog at Brindle of War, http://www.brindle-at-war.net/boerwar.htm (accessed December 25, 2009).

36. Stanley Weintraub, *Silent Night: The Story of the World War I Christmas Truce* (New York: Plume, 2002), 13.

37. Ibid. 25.

38. Ibid., 33.

39. Ibid., 79–80.

40. Ibid. 81.

41. Ibid., 86.

42. Ibid., 113.

43. Ibid., 149.

44. Ibid. 160.

45. Ibid., 169–170.

46. The Deuce of Clubs Book Club, http://www.deuceofclubs.com/books/177xmas.htm (accessed December 25, 2009), 172–173.

47. Williamson Murray, *The Gathering Storm: From World War I to World War II* (Cambridge, MA: Belknap Press of Harvard University Press, 2001), 33.

48. Nicolas Grimal, *A History of Ancient Egypt* (New York: Blackwell Books, 1992), 256–257.

49. Ibid., 256.

50. Ibid., 257.

51. Clarke, "Conflict Termination," 9.

52. Gideon Rose, "How Vietnam Really Ended: Events Abroad—Not Domestic Anti-War Activism—Brought the War to an End," *Slate Online Magazine*, posted Monday, Jan. 22, 2007, http://www.slate.com/id/2158016.

53. Note: The last U.S. serviceman to die in combat in Vietnam, Lt. Col. William B. Nolde, was killed by an artillery shell at An Loc, 60 miles northwest of Saigon, only 11 hours before the truce went into effect.History.com, this day in history presented by Toyota at http://www.history.com/this-day-in-history/paris-peace-accords-signed.

54. Noam Chomsky, "Kosovo Peace Accord" *Z Magazine*, July, 1999, http://www.chomsky.info/articles/199907—.html.

55. Anonymous, "Many Bosnians hope the post-war settlement will give new impetus to revise the Dayton accords along the lines of the solution for Kosovo." Institute for War and Peace Reporting, Kosovo Consequences for Bosnia, March 16, 2010, http://www.iwpr.net/report-news/kosovo-consequences-bosnia.

56. Clarke, "Conflict Termination," 10.

57. CNN.COM/Word, "Bush Calls End to 'Major Combat'" posted May 2, 2003, http://www.cnn.com/2003/WORLD/meast/05/01/sprj.irq.main/.

58. Mark Garrard, "War Termination in the Persian Gulf," *Airpower Journal* (Fall 2001), 1, 4 of the online edition, http://airpower.au.af.mil/airchronicles/apj/apj01/fal101/garrard.html.

59. Ibid., 2, 4.

60. Clarke, "Conflict Termination," 10.

61. Delia K. Cabe, "Nation Building: Shedding Its Isolationist Stance, the United States Begins Reaching Out to Its Global Neighbors" *Kennedy School Bulletin* (Spring 2002), http://www.hks.harvard.edu/ksgpress/bulletin/spring2002/features/nation_building.html.

62. Marin Strmecki, "Stability, Security, Reconstruction, and Rule of Law Capabilities," Adapting America's Security Paradigm and Security Agenda (Washington, DC: National Strategy Information Center, 2010), 22 and 56.

63. "Executive Summary," *Joint Forces Quarterly* 53 (2nd Quarter, 2009), 6.

64. Fareed Zakaria, "The Rise of Illiberal Democracy," *Foreign Affairs* (November/December 1997), http://www.foreignaffairs.com/articles/53577/fareed-zakaria/the-rise-of-illiberal-democracy.

65. Joint Publication 3-07.1: Joint Tactics, Techniques, and Procedures for Foreign Internal Defense (FID) (Washington, DC: Government Printing Office, April 30, 2004), ix.

66. Ali A. Allawi, *The Occupation of Iraq: Winning the War, Losing the Peace* (New Haven, CT: Yale University Press, 2007), 10.

67. Dominic J. Caraccilo and Andrea L. Thompson, *Achieving Victory in Iraq: Countering an Insurgency* (Mechanicsburg, PA: Stackpole Books, 2008), 6.

68. Anthony H. Cordesman, "Shape, Clear, Hold, and Build: The Uncertain Lessons of the Afghan and Iraq Wars," *Center for Strategic & International Studies* (September 23, 2009), http://csis.org/publication/shape-clear-hold-and-build-uncertain-lessons-afghan-iraq-wars.

69. Defense Secretary Robert M. Gates, Landon Lecture (Kansas State University), Remarks as Delivered by Defense Secretary in Manhattan, Kansas, Monday, November 26, 2007 at http://www.defense.gov/speeches/speech.aspx?speechid=1199, 3.

70. Ibid.

71. "The National Security Strategy of the United States of America," Office of the President of the United States: White House Documents and Publications, March 2006, http://www.iwar.org.uk/military/resources/nss-2006/index.htm.

72. Frank G. Hoffman, "Hybrid Warfare and Challenges," *Joint Forces Quarterly* 52 (1st Quarter, 2009), 34.

73. Ibid.

74. "Perspectives on Political and Social Regional Stability Impacted by Global Crises," *A Social Science Context* (January 2010), 4–5.

CHAPTER 2

1. General Ray Odierno, discussion on January 14, 2010.

2. Carl von Clausewitz, *On War*, eds. Michael Howard and Peter Paret (Princeton NJ: Princeton University Press, 1976).

3. Ibid., 90–99.

4. The U.S. Department of State HomePage, http://www.state.gov/s/crs/66427.html.

5. Bruce C. Bade, *War Termination: Why Don't We Plan for It?* Carlisle Barracks, PA: U.S. National War College, 1994, 9.

6. William Flavin, "Planning for Conflict Termination and Post-Conflict Success," *Parameters* (Autumn 2003), 96.

7. U.S. Army HomePage, http://www.army.mil/-news/2009/11/30/31114-ramping-up-to-face-the-challenge-of-irregular-warfare/

8. Kenneth M. Pollack, and Irena L. Sargsyan, "The Other Side of the COIN: Perils of Premature Evacuation from Iraq," *The Washington Quarterly*, March 29, 2010, http://www.twq.com/10april/docs/10apr_PollackSargsyan.pdf.

9. U.S. Department of State HomePage, http://www.state.gov/s/crs/66427.html.

10. Paul R. Pillar, *Negotiating Peace* (Princeton, NJ: Princeton University Press, 1983), 246.

11. Bade, "War Termination," 16.

12. Robert R. Soucy II, Kevin A. Shwedo, and John S. Haven II, "War Termination and Joint Planning," *Joint Forces Quarterly* (Summer 1995), 98.

13. Ibid., 97.

14. Bade, "War Termination," 9.

15. U.S. Department of State HomePage, http://www.state.gov/s/crs/66427.html.

16. Harry R. Yager, Strategic *Theory for the 21st Century: The Little Book on Big Strategy* (Carlisle, PA: Strategic Studies Institute, February 2006), v.

17. Frank G. Hoffman, "Hybrid Warfare and Challenges," *Joint Forces Quarterly* 52 (1st Quarter, 2009), 35.

18. U.S. Department of State HomePage, http://www.state.gov/s/crs/66427.html.

19. Janine Davidson, "Principles of Modern American Counterinsurgency: Evolution and Debate," *Brookings Counterinsurgency and Pakistan Paper Series* (Washington, DC: The Brookings Institution, June 9, 2009).

20. Joseph S. Nye, *Soft Power: The Means of Success in World Politics* (New York: Public Affairs, 2004), 2–4.

21. Michael Findlay and Gary Luck, "Interagency, Intergovernmental, Nongovernmental and Private Sector Coordination," A Joint Force Operational Perspective, Joint Forces Command, Focus Paper #3 (February 2009), 2.

22. Lawrence A. Yates, "The U.S. Military's Experience in Stability Operations, 1789–2005," Global War on Terrorism Occasional Paper 15 (Fort Leavenworth, KS: Combat Studies Institute Press, 2005), 2.

23. U.S. Department of State HomePage, http://www.state.gov/s/crs/66427.html.

24. This list is taken from FM 3-07, *Stability Operations and Support Operations* (Washington, DC: Headquarters, Department of the Army, February 2003).

25. Joint Staff, J-7, *Joint Publication 1-02, Department of Defense Dictionary and Associated Terms* (Washington, DC: U.S. Joint Staff, November 30, 2004), 509.

26. Basil H. Liddell Hart, *Strategy* (New York: Meridian, 1991), 353.

27. John Schwanz, "War Termination: The Application of Operational Art to Negotiating Peace," Final report, Department of Joint Military Operations (Naval War College: Newport, RI; 1996), 3.

28. Stephen J. Cimbala, "The Endgame and War," in *Conflict Termination and Military Strategy: Coercion, Persuasion, and War*, eds. Stephen J. Cimbala and Keith A. Dunn, 11 (Boulder, CO: Westview Press, 1987).

29. John R. Boule, "Operational Planning and Conflict Termination," *Joint Forces Quarterly* (Autumn/Winter 2001–2002), 98.

30. U.S. Joint Chiefs of Staff, *Doctrine for Joint Operations*. Joint Pub 3-0 (Washington, DC: 1995): I-9–I-10.

31. U.S. Army, *Field Manual (FM) 3-0, Operations* (Washington, DC: Department of the Army, June 1, 2001), 6–21.

32. Michael C. Griffith, "War Termination: Theory, Doctrine, and Practice," Unpublished Research Paper, U.S. Army Command and Staff College (Fort Leavenworth, KS: 1992), 8.

33. Susan E. Strednansky, "Balancing the Trinity, The Fine Art of Conflict Termination, Maxwell Air Force Base: Alabama, June 1995" 4.

34. Ibid., 5.

35. Ibid., 6.

36. U.S. Department of State HomePage, http://www.state.gov/s/crs/66427.html.

37. Hoffman, "Hybrid Warfare and Challenges," 34.

38. "Forging a New Shield," Project on National Security Reform (Arlington, VA: The Center for the Study of the Presidency, November 2008), vi–x.

39. Bade, "War Termination," 3.

40. Andrew J. Bacevich, *The Limits of Power: The End of American Exceptionalism* (New York: Metropolitan Books, 2008), 112. Max Boot, "What's Next? The Foreign Policy Agenda Beyond Iraq," *Weekly Standard* (May 5, 2003); Thomas Donnelly, "The Underpinnings of the Bush Doctrine, *AEI National Security Outlook* (January 31, 2003); Frederick W. Kagan, "The Korean Parallel: Is It June 1950 All Over Again?" *Weekly Standard* (October 8, 2001).

41. Steve Coll, "The Future of Soldiering," Think Tank: Online Only, *The New Yorker* (June 1, 2009), http://www.newyorker.com/online/blogs/stevecoll/2009/06/the-future-of-Soldiering.html.

42. Defense Secretary Robert M. Gates, Landon Lecture (Kansas State University), Remarks as Delivered by Defense Secretary in Manhattan, Kansas, Monday, November 26, 2007 at http://www.defense.gov/speeches/speech.aspx?speechid=1199, 3–4.

43. Hart, *Strategy*, 333.

44. Ben D. Mor, "Public Diplomacy in Grand Strategy," *Foreign Policy Analysis*, 2006, 158.

45. Barry R. Posen, and Andrew L. Ross, "Competing Visions for U.S. Grand Strategy," *International Security* (Winter 1997), 5–53.

46. At the time of this writing the Obama administration is reportedly attempted to define a new NSS. Ideas, such as removing the terms like "Islamic radicalism" from the 2006 version and using a new version to emphasize that the United States does not view Muslim nations through the lens of terrorism, counterterrorism officials say are reported key parts of the way the United States looks at strategic involvement in the future. The change would be a significant shift in the NSS, a document that previously outlined the Bush Doctrine of preventive war. It currently states, "The struggle against militant Islamic radicalism is the great ideological conflict of the early years of the 21st century."

47. Mor, "Public Diplomacy," 159.

48. Paul Kennedy, "Grand Strategy in War and Peace: Toward a Broader Definition," *Grand Strategies in War and Peace,* ed. Paul Kennedy (New Haven, CT: Yale University Press, 1991) and Mor, "Public Diplomacy," 159.

49. Edward N. Luttwak, comment on "Political Strategies for Coercive Diplomacy and Limited War" by Alvin H. Bernstein, in *Political Warfare and Psychological Operations Rethinking the U.S. Approach,* ed. Frank R. Barnett and Carnes Lord (Washington, DC: National Defense University Press 1988).

50. Mor, "Public Diplomacy," 157–176.

51. Jeffrey Record, "Exit Strategy Delusions," *Parameters* (Winter 2001), http://smallwarsjournal.com/reference/transition.php.

52. U.S. Department of Defense, Report of the Defense Science Board Task Force on Strategic Communications, Washington, DC, Office of the Under Secretary of Defense for Acquisition, Technology, and Logistics, September 2004, 2, http://www.acq.osd.mil/dsb/reports/2004-09-Strategic_Communication.pdf. (Note: This report is a product of the Defense Science Board (DSB). The DSB is a Federal Advisory Committee established to provide independent advice to the Secretary of Defense. Statements, opinions, conclusions, and recommendations in this report do not necessarily represent the official position of the Department of Defense).

53. Ibid., 1–3.

54. Helle Dale, "Public Diplomacy and Strategic Communications Review: Key Issues for Congressional Oversight," The Heritage Foundation's *This Week on Public Diplomacy and National Security Council,* March 22, 2010, http://www.heritage.org/Research/Reports/2010/03/Public-Diplomacy-and-Strategic-Communications-Review-Key-Issues-for-Congressional-Oversight.

55. Jarol B. Manheim, "Strategic Public Diplomacy: Managing Kuwait's Image during the Gulf Conflict," in *Taken by Storm: The Media, Public Opinion, and U.S. Foreign Policy in the Gulf War,* eds. W. Lance Bennett and David L. Paletz, 4 (Chicago: University of Chicago Press: 1994).

56. Hans Tuch, *Communicating with the World: U.S. Public Diplomacy Overseas* (New York: St. Martin's Press: 1990), 3.

57. Abeer Bassiouny Arafa Ali Radwan, "Public Diplomacy and the Case of "Flotilla," AmericanDiplomacy.Org, September 6, 2010, http://www.unc.edu/depts/diplomat/item/2010/0912/oped/op_radwan.html.

58. Tuch, *"Communicating with the World,"* 3.

59. Business for Diplomatic Action, *America's Role in the World, A Business Perspective on Public Diplomacy* (New York: Author, October 2007), 3–4.

60. Defense Secretary Robert M. Gates, Landon Lecture (Kansas State University), Remarks as Delivered by Defense Secretary in Manhattan, Kansas, Monday, November 26, 2007 at http://www.defense.gov/speeches/speech.aspx?speechid=1199, 5.

61. Nathan Hodge, "U.S. Fighting Off White Phosphorus Allegations, Again" Danger Room: What's Next in National Security, *Wired,* May 11, 2009, http://www.wired.com/dangerroom/2009/05/halt-to-afghan-airstrikes-not-too-likely-says-obama-advisor/.

62. Noah Shachtman, "Ex-Air Force Chief: Recruit Bloggers to Wage Afghan Info War," Danger Room: What's Next in National Security, *Wired,* May 13, 2009, http://www.wired.com/dangerroom/2009/05/ex-air-force-chief-recruit-bloggers-to-wage-afghan-info-war/.

63. Shachtman, Noah, "Ex-Air Force Chief: Recruit Bloggers to Wage Afghan Info War," Danger Room: What's Next in National Security, *Wired,* May 13, 2009 found at http://www.wired.com/dangerroom/2009/05/ex-air-force-chief-recruit-bloggers-to-wage-afghan-info-war/.

64. Noah Shachtman, "Info Wars: Pentagon Could Learn From Obama, Israel," Danger Room: What's Next in National Security, *Wired,* February 25, 2009 found at http://www.wired.com/dangerroom/2009/02/info-war-pentag/.

65. Noah Shachtman, "New Army Rules Could Kill G.I. Blogs (Maybe E-mail, Too)," Danger Room: What's Next in National Security, *Wired,* May 2, 2007, http://www.wired.com/dangerroom/2007/05/new_army_rules_/.

CHAPTER 3

1. Michael E. Brown, J. Coté, R. Owen, Sean M. Lynn-Jones, and Steven E. Miller, eds. *New Global Dangers: Changing Dimensions of International Security* (Cambridge, MA: MIT Press, 2004), 39.

2. Fred C. Iklé, *Every War Must End* (New York: Columbia University Press, 1971), 17–18.

3. Lawrence A. Yates, *The US Military's Experience in Stability Operations, 1789–2005: Global War on Terrorism Occasional Paper 15* (Fort Leavenworth, KS: Combat Studies Institute Press, 2005), 74.

4. Ibid.

5. Ibid., 77.

6. George Bush, "Address to the Nation Announcing United States Military Action in Panama," *Public Papers of the Presidents, 1989,* Washington DC: US Government Printing Office, 1:1723.

7. Ibid.

8. Jessica Wayne, "Operation Just Cause: A Historical Analysis" *Council on Human Affairs,* http://www.coha.org/operation-just-cause-a-historical-analysis/.

9. Yates, *The US Military's Experience,* 92–93.

10. Jeffrey Record, "Exit Strategy Delusions," *Parameters,* Winter 2001, http://smallwarsjournal.com/reference/transition.php.

11. It is not lost on the author that when conditions on the ground don't change then a plan can be easily followed and not open to scrutiny. Since the conditions and environment provided the Coalition the ability to follow the premade plan, then this is arguably a simplistic model to follow because it is clear that those plans requiring an adjustment based on changing conditions are much more difficult and complex in nature.

12. Jeffrey Record, *Making War, Thinking History: Munich, Vietnam, and Presidential Uses of Force from Korea to Kosovo* (Annapolis, MD: The Naval Institute Press, 2002), 105.

13. Dominic J. Caraccilo, "Terminating the Ground War in the Persian Gulf: A Clausewitzian Examination" (Arlington, VA: Association of the United States Army, Institute of Land Warfare, 1997), 15.

14. President Bush Address to the Nation, March 17, 2003, http://www.whitehouse.gov/news/releases/2003/03/20030317-7.html.

15. Steven J. Cimbala and Sidney R. Waldman, eds. *Controlling and Ending Conflict* (Westport, CT: Greenwood Press, 1992), 3.

16. In war, termination design is driven in part by the nature of the war itself. Wars over territorial disputes or economic advantages tend to be interest-based and lend themselves to negotiation, persuasion, and coercion. Wars fought in the name of ideology, ethnicity, or religious or cultural primacy tend to be value-based and reflect demands that are seldom negotiable. Often, wars are a result of both value- and interest-based differences. *Joint Publication 3-0*, III-I9 and Frederick J. Ourso, *War Termination: Do Planning Principles Change with the Nature of the War?* (Newport, RI: Naval War College, May 18, 1998), 9.

17. Stephen Biddle, "War Aims and War Termination," Strategic Issue Analysis. Defeating Terrorism. http://www.911investigations.net/IMG/pdf/doc-137.pdf., 4.

18. President George W. Bush's Address Regarding Ultimatum to Iraq, http://www.johnstonsarchive.net/terrorism/bushiraq2.html.

19. Ibid.

20. Christopher Bassford, and Edward J. Villacres, "Reclaiming the Clausewitzian Trinity" *Parameters* (Autumn, 1995), http://www.clausewitz.com/readings/Bassford/Trinity/TRININTR.htm and Carl von Clausewitz, *On War*, ed. and trans. Michael Howard and Peter Paret, 606 (Princeton, NJ: Princeton University Press, 1984).

21. On November 19, 2009, Rep. David Obey, D-Wis., introduced H.R. 4130, the "Share the Sacrifice Act of 2010." It would establish a 1 percent surtax on everyone's federal income tax liability plus an additional percentage on those with a liability over $22,600 (for couples filing jointly), such that revenue from the surtax would pay for the additional cost of fighting the war in Afghanistan. William R. Gruver, "Fog of War: Why Clausewitz Would Not Be Happy with Obama's New Afghanistan Strategy." *The New Republic* (December 8, 2009), http://www.tnr.com/article/world/fog-war.

22. http://en.wikipedia.org/wiki/2003_invasion_of_Iraq#Bush_declares_.22End_of_major_combat_operations.22_.28May_2003.29

23. General Ray Odierno, Testimony to Congress on September 30, 2009.

24. http://www.latinamericanstudies.org/uruguay/tupamaros-uruguay.htm.

25. *FM 3-24: Counterinsurgency, Operations* (Washington, DC: Department of the Army, December, 2006. p. 3-5.

26. William Flavin, "Planning for Conflict Termination and Post-Conflict Success," *Parameters* (Autumn 2003), 107.

CHAPTER 4

1. Susan E. Strednansky, "Balancing the Trinity: The Fine Art of Conflict Termination," Maxwell Airforce Base, Alabama (June 1995), 5.

2. Joseph A. Engelbregcht, Jr., "War Termination: Why Does a State Decide to Stop Fighting?" (Ph.D. Thesis, Ann Arbor: UMI Dissertation Services, 1992), 24–45.

3. Natalia Bajanova, "Assessing the Conclusion and Outcome of the Korean War," Paper presented to the Korean Society Conference on the "Korean War: An Assessment of the Historical Record," July 24–25, 1995 (Washington, DC: The Georgetown University Conference Center), 5.

4. Ibid., 3.

5. Callum A. MacDonald, *Korea: The War before Vietnam* (Oxford University Press, 1986), 116.

6. Tan Edlin, "The Korean War Explained with Termination of War Theories," *Journal of Singapore Armed Forces Journal* 26, no. 4 (Oct.–Dec. 2000), http://www.mindef.gov.sg/safti/pointer/back/journals/2000/Vol26_4/2.htm.

7. Gay M. Hammerman, *Conventional Attrition and Battle Termination Criteria: A Study of War Termination* (Loring, VA: Defense Nuclear Agency Rpt. No. DNA-TR-81-224, August 1982), 11.

8. William Flavin, "Planning for Conflict Termination and Post-Conflict Success," *Parameters* (Autumn 2003), p. 95.

9. Fred Charles Iklé, *Every War Must End*. New York: Columbia University Press, 1971. 6–7.

10. Yitschak Ben Gad, *Politics, Lies and Videotape: 3,000 Questions and Answers on the Mideast Crisis* (New York: SPI Books, 1991), 175

11. http://www.jewishvirtuallibrary.org/jsource/History/Suez_War.html.

12. Edward C. Luck and Stuart Albert, eds. *On the Endings of Wars* (Port Washington, NY: Kennikat Press Corp, 1980), 5.

13. Andreas W. Daum, Lloyd C. Gardner, and Wilfried Mausbach, eds., *America, the Vietnam War, and the World* (Cambridge, UK: Press Syndicate of the University of Cambridge, 2003), 105.

14. Ibid., 118–119.

15. H.R. McMaster, "Graduated Pressure: President Johnson and the Joint Chiefs," *Joint Force Quarterly*, 34 (Spring 2003): 86.

16. Daum, Gardner, and Mausbach, *America, the Vietnam War,* 119.

17. Linda A. Legier-Topp, "War Termination: Setting Conditions for Peace," (U.S. Army War College thesis, U.S. Army War College, Carlisle Barracks, PA), 9.

18. Lawrence A. Yates, *The U.S. Military's Experience in Stability Operations, 1789–2005: Global War on Terrorism Occasional Paper 15* (Fort Leavenworth, KS: Combat Studies Institute Press, 2005), 82–83.

19. Susan E. Strednansky, "Balancing the Trinity," 12–15.

20. John F. Harris, "Clinton Will Keep Troops in Bosnia," *Washington Post*, December 19, 1997, A01.

21. Wesley Clark, *Waging Modern War* (New York: Perseus Books, 2000), 422–426.

22. Douglas J. Feith, *War and Decision: Inside the Pentagon at the Dawn of the War on Terrorism* (New York: HarperCollins, 2008), 7.

23. Max Boot, "Can We Win in Afghanistan?" *Philadelphia Inquirer* (February 22, 2009), http://www.commentarymagazine.com/viewarticle.cfm/we-can-win-in-afghanistan-15074.

24. Dexter Filkins, "Afghan Civilian Deaths Rose 40% in 2008," *New York Times* (February 17, 2009), http://www.nytimes.com/2009/02/18/world/asia/18afghan.html.

25. Pollack, Kenneth, "The Other Side of the COIN: Perils of Premature Evacuation from Iraq," *The Washington Quarterly* 33, no. 2 (April 2010), 17–32.

26. *FM 3-2, Counterinsurgency Operations*, 1–147.

27. Attributed to Lieutenant General David Rodriguez, Commander of the ISAF Joint Command, Operation Enduring Freedom, January 2010.

28. Attributed to General Stanley McChrystal, Commander of the International Security Assistance Force, Operation Enduring Freedom, January 2010.

29. Delia K. Cabe, "Nation Building: Shedding Its Isolationist Stance, the United States Begins Reaching Out to Its Global Neighbors," *Kennedy School Bulletin* (Spring 2002), http://www.hks.harvard.edu/ksgpress/bulletin/spring2002/features/nation_building.html.

30. General Stanley McChrystal, NPR, June 19, 2009, Morning Edition (NPR) 7:10 A.M.

31. Kori Schacke, "How Not to Lose Afghanistan," *New York Times*, January 26, 2009, http://roomfordebate.blogs.nytimes.com/2009/01/26/how-not-to-lose-afghanistan/.

32. David S. Cloud, "Pentagon Preps for Years in Afghanistan" (April 21, 2009), www.politico.com.

CHAPTER 5

1. Representative Marsha Blackburn (Republican, Tennessee) stated this at the Normandy Dining Inn at Fort Campbell, Kentucky, in November 2004.

2. Richard A. Lacquement, Jr., "Integrating Civilian and Military Activities," *Parameters* 40, no. 1 (Spring 2010), 20.

3. John R. Broule, "Operational Planning and Conflict Termination" *Joint Forces Quarterly* (Autumn/Winter 2001–02), 97.

4. Defense Secretary Robert M. Gates, Landon Lecture (Kansas State University), Remarks as Delivered by Defense Secretary in Manhattan, Kansas, Monday, November 26, 2007 at http://www.defense.gov/speeches/speech.aspx?speechid=1199, 5.

5. Joseph. S. Nye, *Soft Power: The Means of Success in World Politics* (New York: Public Affairs, 2004), 1. Note: "Knowledge is power" is a historical reference reportedly spoken first by Seneca that has progressed into "Information is power" as technology evolved.

6. http://www.emachiavelli.com/Machiavelli%20on%20power.htm.

7. Niccolo Machiavelli, *The Prince*, CreateSpace, 2010, 56.

8. Ibid., 63.

9. Ibid., 60.

10. http://www.emachiavelli.com/Machiavelli%20on%20power.htm.

11. http://www.state.gov/s/crs//.

12. Secretary of State Hillary Rodham Clinton, at the Global Press Conference, Foreign Press, Center, Washington, DC, May 19, 2009.

13. Ibid.

14. http://www.crs.state.gov/index.cfm?fuseaction=public.display&shortcut=CRPF.

15. Remarks by Henrietta H. Fore, Administrator, USAID, and Director of U.S. Foreign Assistance. The Future of Foreign Assistance, Center for U.S. Global

Engagement—2008 Washington Conference Election 2008: The Global Impact. The Mayflower Hotel, Grand Ballroom, Washington, DC., Opening Plenary Remarks; July 15, 2008.

16. Gates, p. 4.

17. Paul Collier, *The Bottom Billion: Why the Poorest Countries Are Failing and What Can Be Done about It* (New York: Oxford University Press, 2008), x.

18. As found on the MCC home page at http://www.mcc.gov/mcc/about/index.shtml.

19. J. Brian Atwood, M. Peter McPherson, and Andrew Natsios, "Arrested Development: Making Foreign Aid a More Effective Tool," *Foreign Affairs* (November/December 2008), 4–5.

20. Secretary of Defense Memorandum, "Options for Remodeling Security Sector Assistance Authorities," December 15, 2009, 1.

21. Section 1206 of the National Defense Authorization Act (NDAA) for Fiscal Year 2006 provides the Secretary of Defense with authority to train and equip foreign military and foreign maritime security forces. The Department of Defense (DoD) values this authority as an important tool to train and equip military partners. Funds may be obligated only with the concurrence of the Secretary of State. Thus far, the DoD has used Section 1206 authority primarily to provide counterterrorism support. This authority expires in FY2011. Section 1206 of the FY2006 NDAA, P.L. 109-163, as amended, provides the Secretary of Defense with a new authority to train and equip foreign military forces and foreign maritime security forces. Section 1206 is the first major DoD authority to be used expressly for the purpose of training the national military forces of foreign countries. Generally, DoD has trained and equipped foreign military forces through State Department programs. The George W. Bush Administration requested this "Global Train and Equip" authority because DoD viewed the planning and implementation processes under which similar State Department security assistance is provided as too slow and cumbersome. Section 1206 provides the Secretary of Defense with authority to train and equip foreign military forces for two purposes. One is to enable foreign military forces, as well as foreign maritime security forces, to perform counterterrorism (CT) operations. Nearly all Section 1206 assistance to date has been CT training and equipment (T&E). Most T&E has been provided by contractors, according to DoD officials. The other purpose is to enable foreign military forces to participate in or to support military and stability operations in which U.S. armed forces are participating. (DoD does not use Section 1206 authority for operations in Iraq and Afghanistan, however, according to DoD officials.) In its May 2009 budget submission for DoD, the Obama Administration requested a $400 million appropriation for Section 1206 spending. This is $50 million above the authorized spending limit of $350 million. According to the DoD FY2010 Budget Request Summary Justification Document accompanying the request, U.S. military "Combatant Commanders consider this [Section 1206] program . . . as the single most important tool for the Department to shape the environment and counter terrorism." According to that document, the Section 1206 program is important because it allows the United States to train and equip foreign military forces to respond to "urgent and emergent threats," and because it "provides opportunities to solve problems before they become crises." In the wake of the September 11, 2001, terrorist attacks on the United States, some DoD officials

sought a means to increase U.S. support to foreign military and security forces to disrupt terrorist networks. Although "train and equip" authority had resided with the State Department since 1961, DoD submitted proposed legislation to Congress in early 2005 for authority and appropriations to train and equip foreign forces. As submitted to Congress, the DoD-proposed legislation differed in several important respects from the legislation that was eventually passed. Nina M. Serafino, "Section 1206 of the National Defense Authorization Act for FY2006: A Fact Sheet on Department of Defense Authority to Train and Equip Foreign Military Forces Specialist in International Security Affairs," Congressional Research Service, Washington, DC, September 28, 2009. State Department programs under which foreign military forces are trained are the International Military Education and Training (IMET) and the Expanded IMET (E-IMET) programs. Equipment is provided to foreign governments through the State Department Foreign Military Sales/Foreign Military Financing (FMS/FMF) programs. According to DoD, this "traditional security assistance takes three to four years from concept to execution," while "Global Train and Equip authority allows a response to emergent threats or opportunities in six months or less." U.S. Department of Defense, *Fiscal Year 2009 Budget Request Summary Justification*, February 4, 2008, 103. Hereafter referred to as *FY2009 DOD Summary Justification*. U.S. Department of Defense, *Fiscal Year 2010 Budget Request Summary Justification*, May 2009, pp. 1–13.

22. Section 1207 of the National Defense Authorization Act for FY2006 (P.L. 109-163) provides authority for DoD to transfer to the State Department up to $100 million in defense articles, services, training, or other support for reconstruction, stabilization, and security activities in foreign countries. This authority expires at the end of FY2008. The amounts that have been transferred thus far were $10 million in FY2006 and $99.5 million in FY2007. Section 1207 of the National Defense Authorization Act for FY2006 (FY2006 NDAA, P.L. 109-163) provides authority for DoD to transfer to the State Department up to $100 million in defense articles, services, training, or other support in FY2006 and again in FY2007 to use for reconstruction, stabilization, and security activities in foreign countries. This authority was extended through FY2008 by Section 1210 of the FY2008 NDAA (P.L. 110-181), which amended the original legislation. Section 1207 authority has been used to fund activities of the State Department's Office of the Coordinator for Reconstruction and Stabilization (S/CRS) and activities implemented by other agencies that are coordinated by S/CRS. Operations and maintenance funds from the three military services and from the defense-wide account have been tapped for this purpose, although the legislation does not specify a funding source. Nina M. Serafino, "Section 1207 of the National Defense Authorization Act for FY2006: Security and Stabilization Assistance: A Fact Sheet," Congressional Research Service, Washington, DC, May 7, 2008.

23. Department of State Memorandum, "Options for Remodeling Security Sector Assistance Authorities," Washington, DC, December 15, 2009.

24. http://www.washingtonpost.com/wp-srv/nation/documents/Gates _to_Clinton_121509.pdf.

CHAPTER 6

1. Soft power is about obtaining desired outcomes through attraction rather than threat or coercion (Nye 2004).

2. Bruce B.G. Clarke, *Conflict Termination: A Rational Model* (Carlisle Barracks, PA: US Army War College, Strategic Studies Institute, 1992), 12.

3. Chantel de Jonge Oudraat, and P.J. Simmons, eds., "From Accord to Action," in *Managing Global Issues: Lessons Learned*, eds. Chantel de Jonge Oudraat and P.J. Simmons (Washington, DC: Carnegie Endowment for International Peace, 2001), 722–723.

4. General Ray Odierno in a discussion with the Kurdistan Regional Minister of Defense and Peshmerga Commander on January 6, 2010.

5. Milan N. Vego, "Systems versus Classical Approach to Warfare." *Joint Forces Quarterly* 52 (1st Quarter 2009), 41.

6. David L. Grange, and Scott Swanson, "Confronting Irregular Challenges in the 21st Century," *Irregular Concept Series: Whole of World Collaboration* (March 10. 2009), 1–3.

7. Thomas P. Barnett, "The Pentagon's New Map: It Explains Why We're Going to War and Why We'll Keep Going to War." *Esquire* (March 2003), 1.

CHAPTER 7

1. Bruce B.G. Clarke, *Conflict Termination: A Rational Model* (Carlisle Barracks, PA: U.S. Army War College, 1992), 30.

2. Ibid., 30, 32.

3. Broule, John R. "Operational Planning and Conflict Termination," *Joint Forces Quarterly* (Autumn/Winter 2001–2002), 99–100.

4. The idea is not a new one. Gen Matthew Ridgway unsuccessfully attempted to acquire expertise from the State Department during his peace talks in Korea. Joseph McMillan, "Talking to the Enemy: Negotiations in Wartime," *Comparative Strategy* 11 (1992), 452.

5. George Bush and Brent Scowcroft, *A World Transformed* (New York: Alfred A. Knopf, 1998), 484.

6. Colin S. Gray, *Modern Strategy* (Oxford: Oxford University Press, 1999), p. 361.

7. Frederick J. Ourso, *War Termination: Do Planning Principles Change with the Nature of the War* (Newport, RI: Naval War College, May 18, 1998), 14.

8. Clarke, "Operational Planning," 9–10.

9. Strednansky, Susan E. "Balancing the Trinity: The Fine Art of Conflict Termination," Maxwell Airforce Base: Alabama, June 1995, 8.

EPILOGUE

1. R. Broule, "Operational Planning and Conflict Termination," *Joint Forces Quarterly* (Autumn/Winter 2001–2002), 97.

2. Julius Stone, "International Conflict Resolution," in *International Encyclopedia of the Social Sciences*, reprint edition 1972.

3. James W. Reed, "Should Deterrence Fail: War Termination in Campaign Planning," *Parameters* (Summer 1993), 41–52.

4. Fred Iklé, *Every War Must End* (New York: Columbia University Press, 1971), 85.

5. Fareed Zakaria, "Victory in Iraq: How We Got Here Is a Matter for History. But the Democratic Ideal Is Still within Reach." *Newsweek* (June 15, 2009), http://www.newsweek.com/id/200858.

Bibliography

"Active Component Stress on the Force." Alexandria, VA: U.S. Army, Deputy Chief of Staff for Operations, Training and Analysis Division, July 2009.

Agreement on Ending the War and Restoring Peace in Vietnam (signed in Paris and entered into force January 17, 1973), http://www.mtholyoke.edu/acad/intrel/vietnam/treaty.html.

Allawi, Ali A. *The Occupation of Iraq: Winning the War, Losing the Peace,* New Haven, CT: Yale University Press, 2007.

Allison, Graham T. Albert Carnesale, and Joseph S. Nye Jr. *Hawks, Doves, and Owls: An Agenda for Avoiding Nuclear War.* New York: W. W. Norton, 1985.

Allison, Graham T., and William L. Ury, with Bruce J. Allyn. *Windows of Opportunity: From Cold War to Peaceful Competition in U.S.-Soviet Relations,* Cambridge, MA: Ballinger Publishing, 1989.

Anonymous. "Many Bosnians Hope the Post-War Settlement Will Give New Impetus to Revise the Dayton Accords along the Lines of the Solution for Kosovo." Institute for War and Peace Reporting, Kosovo Consequences for Bosnia, March 16, 2010, http://www.iwpr.net/report-news/kosovo-consequences-bosnia.

"The Armistice." *The War to End All Wars.* Found at www.FirstWorldWar.com, May 1, 2004.

Atkinson, Rick. *Crusade,* New York: Mariner, 1994.

Atwood, J. Brian, M. Peter McPherson, and Andrew Natsios. "Arrested Development: Making Foreign Aid a More Effective Tool." *Foreign Affairs* (November/December 2008).

Bacevich, Andrew J. *The Limits of Power: The End of American Exceptionalism.* New York: Metropolitan Books, 2008.

Bade, Bruce C. *War Termination: Why Don't We Plan for It?* Carlisle Barracks, PA: U.S. National War College, 1994.

Bailey, Sydney D. "Cease-Fires, Truces, and Armistices in the Practices of the UN Security Council," *American Journal of International Law* 71 (1971).

Bajanova, Natalia. "Assessing the Conclusion and Outcome of the Korean War," Paper presented to the Korean Society Conference on the Korean War: An Assessment of the Historical Record, July 24–25, 1995, Washington, DC: The Georgetown University Conference Center.

Ball, Desmond. *Can Nuclear War Be Controlled? Adelphi Papers* 169. London: International Institute for Strategic Studies, 1981.

Barnett, Thomas P. "The Pentagon's New Map: It Explains Why We're Going to War and Why We'll Keep Going to War." *Esquire* (March 2003).

Bassford, Christopher, and Edward J. Villacres. "Reclaiming the Clausewitzian Trinity," *Parameters* (Autumn, 1995), http://www.clausewitz.com/readings/Bassford/Trinity/TRININTR.htm.

Bathurst, Robert B. *Some Problems in Soviet-American War Termination: Cross-Cultural Asymmetries.* NPS-56-88-028, Monterey, CA: U.S. Naval Postgraduate School, September 1988.

Biddle, Stephen. "War Aims and War Termination," Strategic Issue Analysis. Defeating Terrorism, Strategic Studies Institute, http://www.911investigations.net/IMG/pdf/doc-137.pdf.

Blainey, Geoffrey. *The Causes of War.* New York: Free Press, 1973.

Blair, Bruce G. *Strategic Command and Control: Redefining the Nuclear Threat.* Washington, DC: Brookings Institution, 1985.

Blight, James G., and David A. Welch. *On the Brink: Americans and Soviets Reexamine the Cuban Missile Crisis.* New York: Hill and Wang, 1989.

Boas, Gideon, James L. Bischoff, and Natalie L. Reid. *Elements of Crimes under International Law,* International Criminal Law Practitioner Library, Vol. 2. New York: Cambridge University Press, 2009.

Boot, Max. "Can We Win in Afghanistan?" *Philadelphia Inquirer* (February 22, 2009), http://www.commentarymagazine.com/viewarticle.cfm/we-can-win-in-afghanistan-15074.

Boot, Max. "What's Next? The Foreign Policy Agenda Beyond Iraq," *Weekly Standard,* May 5, 2003.

Borden, William L. *There Will Be No Time.* New York: Macmillan, 1946.

British Army Field Manual, Countering Insurgency (Volume 1, Part 10): Army Code 71876, October 2009.

Brodie, Bernard, ed. *The Absolute Weapon.* New York: Harcourt Brace, 1946.

Broule, John R. "Operational Planning and Conflict Termination" *Joint Forces Quarterly* (Autumn/Winter 2001–02).

Brown, Michael E., J. Coté, R. Owen, Sean M. Lynn-Jones, and Steven E. Miller, eds. *New Global Dangers: Changing Dimensions of International Security.* Cambridge, MA: MIT Press, 2004.

Burden, Matthew Currier. *The Blog of War: Front-line Dispatches from Soldiers in Iraq and Afghanistan.* New York: Simon & Shuster, 2006.

Bush, George. *Public Papers of the Presidents, 1989,* 1:1723, Address to the nation announcing United States military action in Panama.

Bush, George, and Brent Scowcroft. *A World Transformed,* New York: Alfred A. Knopf, 1998.

Business for Diplomatic Action. *America's Role in the World: A Business Perspective on Public Diplomacy.* New York: Author, October 2007.

Cabe, Delia K. "Nation Building: Shedding Its Isolationist Stance, the United States Begins Reaching Out to Its Global Neighbors" *Kennedy School Bulletin* (Spring 2002), http://www.hks.harvard.edu/ksgpress/bulletin/spring2002/features/nation_building.html.

Caraccilo, Dominic J. *Achieving Victory in Iraq: Countering an Insurgency.* Mechanicsburg, PA: Stackpole Books, 2005.

Caraccilo, Dominic J. *Terminating the Ground War in the Persian Gulf: A Clausewitzian Examination.* Arlington, VA: Association of the United States Army, Institute of Land Warfare, 1997.

Caraccilo, Dominic J., and Andrea L. Thompson. *Achieving Victory in Iraq: Countering an Insurgency,* Mechanicsburg, PA; Stackpole Books, 2005.

CBBC. "What Is Remembrance Day?" CBBC Newsround, http://news.bbc.co.uk/cbbcnews/hi/find_out/guides/uk/remembrance_day/newsid_2438000/2438201.stm. Accessed December 27, 2009.

Chomsky, Noam. "Kosovo Peace Accord" *Z Magazine* (July, 1999) http://www.chomsky.info/articles/199907—.htm.

Cimbala, Stephen J. *Nuclear Endings: Stopping War on Time.* Westport, CT: Praeger, 1989.

Cimbala, Stephen J. *Through a Glass Darkly: Looking at Conflict Prevention, Management, and Termination,* Westport, CT: Praeger Publishers, 2001.

Cimbala, Steven J., and Keith A. Dunn, eds. *Conflict Termination and Military Strategy: Coercion, Persuasion, and War.* Boulder, CO: Westview Press, 1987.

Cimbala, Steven J., and Sidney R. Waldman, eds. *Controlling and Ending Conflict.* Westport, CT: Greenwood Press, 1992.

Clark, Wesley, *Waging Modern War.* New York: Perseus Books, 2000.

Clarke, Bruce B.G. *Conflict Termination: A Rational Model.* Carlisle Barracks, PA: U.S. Army War College, 1992.

Clausewitz, Carl von. *On War,* edited and translated by Michael Howard and Peter Paret. Princeton, New Jersey: Princeton University Press, 1976, 1984.

Clinton, Secretary of State Hillary Rodham. The Global Press Conference, Foreign Press, Center, Washington, DC, May 19, 2009.

Cloud, David S. "Pentagon Preps for Years in Afghanistan" (April, 21, 2009), www.politico.com.

CNN. "1949 Armistice." *Middle East, Land of Conflict.* CNN. http://www.cnn.com/SPECIALS/2001/mideast/stories/history.maps/armistice.html. Accessed December 27, 2009.

Coalson, Robert A. "Year after Russia-Georgia War—A New Reality, But Old Relations," *Radio Free Europe, Radio Liberty,* April 29, 2010, http://www.rferl.org/content/A_Year_After_RussiaGeorgia_War__A_New_Reality_But_Old_Relations/1793048.html.

Cockburn, Patrick. "Sadr Calls Six-Month Ceasefire to Prevent Civil War," *The Independent World,* August 30, 2007, http://www.independent.co.uk/news/world/middle-east/sadr-calls-sixmonth-ceasefire-to-prevent-civil-war-463540.html.

Coles, David J., Stephen David Heidler, Jeanne T. Heidler, and James M. McPherson, *Encyclopedia of the American Civil War: A Political, Social, and Military History.* New York: W.W. Norton, 2002.

Coll, Steve. "The Future of Soldiering," Think Tank: Online Only. *The New Yorker* (June 1, 2009), http://www.newyorker.com/online/blogs/stevecoll/2009/06/the-future-of-soldiering.html.

Collier, Paul. *The Bottom Billion: Why the Poorest Countries Are Failing and What Can Be Done about It,* New York: Oxford University Press, 2008.

Corbett, Sir Julian. *Principles of Maritime Strategy.* London: Conway Maritime Press, 1911.

Cordesman, Anthony H. "Shape, Clear, Hold, and Build: The Uncertain Lessons of the Afghan and Iraq Wars," *Center for Strategic & International Studies* (September 23, 2009), http://csis.org/publication/shape-clear-hold-and-build-uncertain-lessons-afghan-iraq-wars.

Dale, Helle. "Public Diplomacy and Strategic Communications Review: Key Issues for Congressional Oversight," The Heritage Foundation (March 22, 2010), http://www.heritage.org/Research/Reports/2010/03/Public-Diplomacy-and-Strategic-Communications-Review-Key-Issues-for-Congressional-Oversight.

Daum, Andreas W., Lloyd C. Gardner, and Wilfried Mausbach, eds., *America, the Vietnam War, and the World: Comparative and International Perspectives.* Cambridge, UK: Cambridge University Press, 2003.

Davidson, Janine. "Principles of Modern American Counterinsurgency: Evolution and Debate," *Brookings Counterinsurgency and Pakistan Paper Series.* Washington, DC: The Brookings Institution, June 9, 2009.

Davis, Paul K. "Applying Artificial Intelligence Techniques to Strategic Level Gaming and Simulation." In *Simulation in the Artificial Intelligence Era*, edited by Maurice Elzas et al. Amsterdam: North-Holland, 1986.

Davis, Paul K. *A New Analytic Technique for the Study of Deterrence, Escalation.* Santa Monica, CA: RAND Corporation, 1986.

Davis, Paul K., Steven C. Bankes, and James P. Kahan. *A New Methodology for Modeling National Command Level Decisionmaking in War Games and Simulations.* Santa Monica, CA: RAND Corporation, 1986.

de Jonge Oudraat, Chantel, and P.J. Simmons. "From Accord to Action." In *Managing Global Issues: Lessons Learned,* edited by Chantel de Jonge Oudraat, and P.J. Simmons. Washington, DC: Carnegie Endowment for International Peace, 2001.

Department of State Memorandum. "Options for Remodeling Security Sector Assistance Authorities," Washington, DC, December 15, 2009.

Department of State Memorandum. *Deterrence and the Revolution in Soviet Military Doctrine.* Washington, DC: Brookings Institution, 1990.

Donnelly, Thomas. "The Underpinnings of the Bush Doctrine, *AEI National Security Outlook,* January 31, 2003.

Eckstein, Harry. *Internal War: Problems and Approaches.* New York: The Free Press, 1964.

Edlin, Tan. "The Korean War Explained with Termination of War Theories," *Journal of Singapore Armed Forces Journal* 26, no. 4 (Oct–Dec 2000), http://www.mindef.gov.sg/safti/pointer/back/journals/2000/Vol26_4/2.htm.

Engelbregcht, Joseph A., Jr. "War Termination: Why Does a State Decide to Stop Fighting?" Ph.D. diss., Ann Arbor: UMI Dissertation Services, 1992.

Engelbregcht, Joseph A., Jr."Executive Summary," *Joint Forces Quarterly* 53 (2nd Quarter, 2009).

Feith, Douglas J. *War and Decision: Inside the Pentagon at the Dawn of the War on Terrorism.* New York: HarperCollins, 2008.

Field Manual (FM) 3-0, Operations (Washington, DC: Department of the Army, June 1, 2001).

Field Manual 3-24: Counterinsurgency Operations (Washington, DC: Department of the Army, December, 2006).

Filkins, Dexter. "Afghan Civilian Deaths Rose 40% in 2008." *New York Times* (February 17, 2009), http://www.nytimes.com/2009/02/18/world/asia/18afghan.html.

Findlay, Michael, and Gary Luck. *Interagency, Intergovernmental, Nongovernmental and Private Sector Coordination,* A Joint Force Operational Perspective, Joint Forces Command, Focus Paper #3, February 2009.

Flavin, William. "Planning for Conflict Termination and Post-Conflict Success," *Parameters* (Autumn 2003).

Fore, Henrietta H., Administrator, USAID, and Director of U.S. Foreign Assistance. The Future of Foreign Assistance, Center for U.S. Global Engagement—2008 Washington Conference Election 2008: The Global Impact. Opening Plenary Remarks. The Mayflower Hotel, Washington, DC, July 15, 2008.

"Forging a New Shield." Arlington, VA: Project on National Security Reform, The Center for the Study of the Presidency, November 2008.

Foster, James L., and Garry D. Brewer. *And the Clock Strikes Thirteen: The Termination of War.* Santa Monica, CA: The RAND Corporation, 1976.

Freedman, Lawrence. *The Evolution of Nuclear Strategy.* New York: St. Martin's Press, 1981.

Friedman, George. *The Next 100 Years: A Forecast for the 21st Century,* New York: Anchor Books, 2010.

Gad, Yitschak Ben, *Politics, Lies and Videotape: 3,000 Questions and Answers on the Mideast Crisis* New York: SPI Books, 1991.

Gaddis, John Lewis. *Strategies of Containment.* New York: Oxford University Press, 1982.

Garrard, Mark. "War Termination in the Persian Gulf," *Airpower Journal* (Fall 2001), p. 4 of the online edition, http://airpower.au.af.mil/airchronicles/apj/apj01/fal101/garrard.html.

Garthoff, Raymond L. *Reflections on the Cuban Missile Crisis,* rev. ed. Washington, DC: Brookings Institution, 1989.

Gates, Robert M. Landon Lecture (Kansas State University), Remarks as Delivered by Secretary of Defense, Manhattan, Kansas, Monday, November 26, 2007 at http://www.defense.gov/speeches/speech.aspx?speechid=1199.

Gayvoronskiy, Col.-Gen. F. F., ed. *Evolutsiya voyennogo iskusstva: etapy tendentsii, printsipy (The evolution of Military Art: Stages, Tendencies, Principles).* Moscow: Voyenizdat, 1987.

George, Alexander L., and Richard Smoke. *Deterrence in American Foreign Policy: Theory and Practice.* New York: Columbia University Press, 1974.

Gorbachev, Mikhail. *Perestroika: New Thinking for Our Country and the World.* New York: Harper and Row, 1987.

Gordon, Michael R., and Trainor, Bernard E. *The Generals' War.* New York: Back Bay Books, 1995.

Grange, David L., and Scott Swanson. "Confronting Irregular Challenges in the 21st Century," Irregular Warfare Concept Series: Whole of World Collaboration, March 10, 2009.

Gray, Colin S. *Modern Strategy,* Oxford: Oxford University Press, 1999.

Gray, Colin S. *Nuclear Strategy: The Case for a Theory of Victory.* In *Strategy and Nuclear Deterrence,* edited by Steven E. Miller, 25–57. Princeton, NJ: Princeton University Press, 1984.

Griffith, Michael C. *War Termination: Theory, Doctrine, and Practice,* Unpublished Research Paper. Fort Leavenworth, KS: U.S. Army Command and Staff College, 1992.

Grimal, Nicolas, *A History of Ancient Egypt,* Lake Oswego, OR: Blackwell Books: 1992.

Gruver, William R. "Fog of War: Why Clausewitz Would Not Be Happy with Obama's New Afghanistan Strategy." *The New Republic* (December 8, 2009), http://www.tnr.com/article/world/fog-war.

Hammerman, Gay M. *Conventional Attrition and Battle Termination Criteria.* Washington, DC: Defense Nuclear Agency, DNA-TR-81-224, 1982.

Harkavy, Robert E. *The Lessons of Recent Wars in the Third World:* Vol. 1, *Approaches and Case Studies,* Lexington, MA; Lexington Books, 1985.

Harris, John F. "Clinton Will Keep Troops in Bosnia." *Washington Post,* December 19, 1997.

Hart, Basil H. Liddell. *Strategy,* New York: Praeger, 1967.

Hart, Basil H. Liddell. *Strategy.* New York: Meridian, 1991.

Heintz, Jim. "On War's Anniversary, Georgia, Russia Vie in Media: Georgia, Russia Push for Public Opinion Victory on War's Anniversary." The Associated Press for ABC International, August 6, 2009, http://abcnews.go.com/International/wireStory?id=8266942.

Hessman, James D. "Three Decades of Mission Creep; Loy: The 'Do More with Less' Well Has Run Dry," http://www.navyleague.org/seapower/three_decades_of_mission_creep.htm.

Hodge, Nathan. "U.S. Fighting Off White Phosphorus Allegations, Again" Danger Room: What's Next in National Security, *Wired,* May 11, 2009, http://www.wired.com/dangerroom/2009/05/halt-to-afghan-airstrikes-not-too-likely-says-obama-advisor/.

Hoffman, Frank G. "Hybrid Warfare and Challenges," *Joint Forces Quarterly* 52 (1st Quarter, 2009).

Holloway, David. *The Soviet Union and the Arms Race.* New Haven, CT: Yale University Press, 1984.

Huntington, Samuel P. "American Military Strategy." In *Policy Papers in International Affairs,* 28, 3–17. Institute of International Studies, 1986.

Huntington, Samuel P. *Political Order in Changing Societies.* New Haven, CT: Yale University Press, 1968.

Iklé, Fred Charles. *Every War Must End.* New York: Columbia University Press, 1971.

Jervis, Robert. *The Illogic of American Nuclear Strategy.* Ithaca, NY: Cornell University Press, 1984.

Jervis, Robert, Richard Ned Lebow, and Janice Gross Stein. *Psychology and Deterrence*. Baltimore, MD: Johns Hopkins University Press, 1985.

Joffe, Josef. *The Limited Partnership: Europe, the United States and the Burdens of Alliance*. Cambridge, MA: Ballinger Publishing, 1987.

Joint Publication 1-02, DoD Dictionary of Military and Associated Terms (Amended through October 31, 2009), http://www.dtic.mil/doctrine/dod_dictionary/.

Joint Publication 3.0, Doctrine for Joint Operations. Washington, DC: Government Printing Office, April 30, 2004.

Joint Publication 3-07.1: Joint Tactics, Techniques, and Procedures for Foreign Internal Defense (FID). Washington, DC: Government Printing Office, April 30, 2004.

Joint Staff, J-7, Joint Publication 1-02, Department of Defense Dictionary and Associated Terms., Washington, DC: U.S. Joint Staff, November 30, 2004.

Kagan, Frederick W. "The Korean Parallel: Is It June 1950 All Over Again?" *Weekly Standard*, October 8, 2001.

Kagan, Kimberly. "The Patton of Counterinsurgency." Institute for the Study of War home page at www.understandingwar.org/other-work/patton-counterinsurgency (accessed January 8, 2010.

Kanveskiy, Boris, and Pyotr Shabardin. "The Correlation of Politics, War, and a Nuclear Catastrophe." *International Affairs* 2 (February 1988).

Kecskemeti, Paul. *Strategic Surrender: The Politics of Victory, and Defeat*. Stanford, CA: Stanford University Press, 1958.

Kennedy, Paul. "Grand Strategy in War and Peace: Toward a Broader Definition," *Grand Strategies in War and Peace*, edited by Paul Kennedy. New Haven, CT: Yale University Press, 1991.

Kokoshin, Andrey, and Maj. Gen. Valentin Larionov. *Protivostoyaniya sil obshchego naznacheniya v kontekste obespecheniya strategicheskoye stabil'nosti (Counterpositioning of General Purpose Forces in the Context of Strategic Stability), Mirovaya Ekonomika i Mezhdunarodnyye Otnosheniya* 6 (June 1988): 23–31.

Lacquement, Richard A., Jr. "Integrating Civilian and Military Activities," *Parameters* 40, no. 1 (Spring 2010).

Lebow, Richard Ned. *Nuclear Crisis Management*. Ithaca, NY: Cornell University Press, 1987.

Legier-Topp, Linda A. *War Termination: Setting Conditions for Peace*, Carlisle Barracks, PA: U.S. Army War College thesis, U.S. Army War College, 2009.

Luck, Edward C., and Stuart Albert, eds. *On the Endings of Wars*. Port Washington, NY: Kennikat Press, 1980.

Luttwak, Edward N. Comment (on "Political Strategies for Coercive Diplomacy and Limited War" by Alvin H. Bernstein). In *Political Warfare and Psychological Operations: Rethinking the U.S. Approach*, edited by Frank R. Barnett and Carnes Lord. Washington, DC: National Defense University Press, 1988.

McChrystal, General Stanley. NPR, June 19, 2009, Morning Edition (NPR) 7:10 A.M.

McChrystal, General Stanley. Commander of the ISAF Joint Command, Operation Enduring Freedom; statements made at ISAF Headquarters in January 2010.

MacDonald, Callum A. Korea: *The War before Vietnam*. New York: Oxford University Press, 1986.

McGwire, Michael. *Military Objectives in Soviet Foreign Policy*. Washington, DC: Brookings Institution, 1987.

Machiavelli, Niccolo. *The Prince*, CreateSpace, 2010.

McMaster, H. R. "Graduated Pressure: President Johnson and the Joint Chiefs," *Joint Force Quarterly* 34 (Spring 2003): 86, cited in *Theory of War and Strategy*, Vol. I, AY 2009, Carlisle Barracks, PA: U.S. Army War College, August 2008.

McMillan, Joseph. "Talking to the Enemy: Negotiations in Wartime," *Comparative Strategy* 11, 1992.

McNamara, Robert S. *Blundering into Disaster*. New York: Pantheon Books, 1986.

McNamara, Robert S. *The Meaning of the Nuclear Revolution*. Ithaca, NY: Cornell University Press, 1989.

Manheim, Jarol B. "Strategic Public Diplomacy: Managing Kuwait's Image during the Gulf Conflict." In *Taken by Storm: The Media, Public Opinion, and U.S. Foreign Policy in the Gulf War*, edited by W. Lance Bennett and David L. Paletz. Chicago: University of Chicago Press, 1994.

Marlow, Ann. "The Surge Afghanistan Needs: More Local Security Forces and a Better Constitution Are Keys to Success." *The Wall Street Journal* (February 15, 2009), http://online.wsj.com/article/SB123457513815386695.html.

Mearsheimer, John J. *Conventional Deterrence*. Ithaca, NY: Cornell University Press, 1983.

Meyer, Stephen M. *Soviet Nuclear Operations*. In *Managing Nuclear Operations*, edited by Ashton B. Carter, John D. Steinbruner, and Charles A. Zraket. Washington, DC: Brookings Institution, 1987.

Mor, Ben D., "Public Diplomacy in Grand Strategy," *Foreign Policy Analysis* 2, no. 2 (April 2006), 157–176.

Morley, Alistair. *Historical Analysis of Conflict Termination*. Fornborough, Hants, UK: Policy and Capabilities Studies Department, Defense Science and Technology Laboratory, http://www.ima.org.uk/conflict/papers/Morley.pdf.

Moyar, Mark. *A Question of Command: Counterinsurgency from the Civil War to Iraq*. New Haven, CT: Yale University Press, 2009.

Mueller, John. *Retreat from Doomsday: The Obsolescence of Major War*. New York: Basic Books, 1989.

Murray, Williamson. *The Gathering Storm: From World War I to World War II*, Cambridge, MA: Belknap Press, 2001.

Nye, Joseph S. *Soft Power: The Means of Success in World Politics*, New York: Public Affairs, 2004.

Odierno, Ray. General Testimony to Congress on September 30, 2009.

Odierno, Ray. Discussion with the Kurdistan Regional Minister of Defense and Peshmerga Commander on January 6, 2010.

Odierno, Ray. Discussion on January 14. 2010.

Ogarkov, Marshal N.V. *Vsegda v gotovnosti k zashchite Otechestva (Always in Readiness to Defend the Fatherland)*. Moscow: Voyenizdat, 1982.

Operation Desert Storm: Evaluation of the Air Campaign (Letter Report, 06/12/97, GAO/NSIAD-97-134); Appendix V; http://www.fas.org/man/gao/nsiad97134/app_05.htm.

Ourso, Frederick J. *War Termination: Do Planning Principles Change with the Nature of the War?* Newport, RI: Naval War College, May 18, 1998.

Paret, Peter, ed. *Makers of Modern Strategy*. Princeton, NJ: Princeton University Press, 1986.

"Perspectives on Political and Social Regional Stability Impacted by Global Crises," *A Social Science Context,* January 2010.

Pillar, Paul. *Negotiating Peace: War Termination as a Bargaining Process*. Princeton, NJ: Princeton University Press, 1983.

Pollack, Kenneth. "The Other Side of the COIN: Perils of Premature Evacuation from Iraq," *The Washington Quarterly* (April 2010).

Posen, Barry R., and Andrew L. Ross. "Competing Visions for U.S. Grand Strategy," *International Security* (Winter 1997).

Powell, Colin L. *My American Journey,* New York: Random House, 1995.

Quester, George H. *Deterrence before Hiroshima*. New York: John Wiley and Sons, 1966, http://findarticles.com/p/articles.

Racueils de la Société Internationale de Droit Pénal Militaire et de Droit de la Guerre. Vol. II: Deuxième Congres International, Florence, May 17–20, 1961.

Radwan, Abeer Bassiouny Arafa Ali, "Public Diplomacy and the Case of "Flotilla," AmericanDiplomacy.Org, September 6, 2010, http://www.unc.edu/depts/diplomat/item/2010/0912/oped/op_radwan.html.

The RAND Corporation. *Control, and War Termination*. Santa Monica, CA: The RAND Corporation, 1986.

Raymer, James H. "In Search of Lasting Results: Military War Termination Doctrine." Fort Leavenworth, KS: Military Monographs, School of Advanced Military Studies, United States Army Command and General Staff College, AY 2001–2002.

Record, Jeffrey. "Exit Strategy Delusions," *Parameters* (Winter 2001), http://findarticles.com/p/articles/mi_m0IBR/is_4_31/ai_82064203/?tag=content;col1.

Record, Jeffrey. *Making War, Thinking History: Munich, Vietnam, and Presidential Uses of Force from Korea to Kosovo,* Annapolis, MD: The Naval Institute Press, 2002.

Reed, James W. "Should Deterrence Fail: War Termination in Campaign Planning," *Parameters* (Summer 1993).

"Report of the Defense Science Board Task Force on Strategic Communication." Washington, DC: Office of the Undersecretary of Defense for Acquisition, Technology, and Logistics (September 2004), http://www.acq.osd.mil/dsb/reports/2004-09-Strategic_Communication.pdf.

Rodriguez, Lieutenant General David, Commander of the ISAF Joint Command, Operation Enduring Freedom; statements made at ISAF Headquarters in January 2010.

Rose, Gideon. "How Vietnam Really Ended: Events Abroad—Not Domestic Anti-War Activism—Brought the War to an End," *Slate Online Magazine*, posted Monday, Jan. 22, 2007, at http://www.slate.com/id/2158016.

Schacke, Kori. "How Not to Lose Afghanistan," *New York Times*, January 26, 2009, http://roomfordebate.blogs.nytimes.com/2009/01/26/how-not-to-lose-afghanistan/.

Schaill, Emmett M. "Planning and End State: Has Doctrine Answered the Need?" Fort Leavenworth, KS: School of Advanced Military Studies Monograph, U.S. Army Command and General Staff College, May 21, 1998.

Schelling, Thomas C. *Arms and Influence*. New Haven, CT: Yale University Press, 1966.

Schwanz, John. "War Termination: The Application of Operational Art to Negotiating Peace," Newport: RI.

Secretary of Defense Memorandum. "Options for Remodeling Security Sector Assistance Authorities," December 15, 2009.

"Section 1207 of the National Defense Authorization Act for FY2006: Security and Stabilization Assistance: A Fact Sheet," Congressional Research Service, Washington, DC, May 7, 2008.

Serafino, Nina M. "Section 1207 of the National Defense Authorization Act for FY2006: Security and Stabilization Assistance: A Fact Sheet," Congressional Research Service, Washington, DC, May 7, 2008.

Shachtman, Noah. "Ex-Air Force Chief: Recruit Bloggers to Wage Afghan Info War," Danger Room: What's Next in National Security, *Wired*, May 13, 2009, http://www.wired.com/dangerroom/2009/05/ex-air-force-chief-recruit-bloggers-to-wage-afghan-info-war/.

Shachtman, Noah. "Info Wars: Pentagon Could Learn From Obama, Israel," Danger Room: What's Next in National Security, *Wired*, February 25, 2009, http://www.wired.com/dangerroom/2009/02/info-war-pentag/.

Shachtman, Noah. "New Army Rules Could Kill G.I. Blogs (Maybe E-mail, Too)," Danger Room: What's Next in National Security. *Wired*, May 2, 2007, http://www.wired.com/dangerroom/2007/05/new_army_rules_/.

Small, Melvin, and J. David Singer. *Resort to Arms: International and Civil Wars, 1816–1980*. Beverly Hills, CA: Sage Publications, 1982.

Smith, David S., ed. *From War to Peace*. New York: Columbia University, 1974.

Sokolovsky, V. D., ed. *Voyennaya strategiya(Military Strategy)*. Moscow: Voyenizdat, 1962.

Soucy, Robert R. II, Kevin A. Shwedo, and John S. Haven, II. "War Termination and Joint Planning," *Joint Forces Quarterly* (Summer, 1995).

Staudenmaier, William O. "Conflict Termination in the Nuclear Era." In *Conflict Termination and Military Strategy: Coercion, Persuasion and War*, edited by Stephen J. Cimbala and Keith A. Dunn. Boulder, CO: Westview Press, 1987.

Staudenmaier, William O. "Conflict Termination in the Third World: Theory and Practice." In *The Lessons of Recent Wars in the Third World: Volume 1, Approaches and Case Studies,* edited by Robert E. Harkavy. Lexington, MA: Lexington Books, 1985.

Stone, Julius. "International Conflict Resolution." In *International Encyclopedia of the Social Sciences*, reprint edition 1972.

Stone, Julius, ed. *The Strategic Imperative: New Policies for National Security*. Cambridge, MA: Ballinger Publishing, 1982.

Strednansky, Susan E. "Balancing the Trinity: The Fine Art of Conflict Termination," Maxwell Airforce Base: Alabama, June 1995.

Strmecki, Marin. "Stability, Security, Reconstruction, and Rule of Law Capabilities," Adapting America's Security Paradigm and Security Agenda. Washington, DC: National Strategy Information Center, 2010,

Strmecki, Marin. *Studying First Strike Stability with Knowledge Based Models of Human Decisionmaking*. Santa Monica, CA: RAND Corporation, 1989.

Summers, Harry G., Jr. *On Strategy: The Vietnam War in Context*. Carlisle Barracks, PA: U.S. Army War College, 1981.

Sun, Tzu. *The Art of War*. Translated by Samuel B. Griffith. Oxford, UK: Oxford University Press, 1963.

Tritten, James J. "Are Nuclear and Nonnuclear War Related?" *Journal of Strategic Studies* 2, no. 3 (September 1988), 365–373.

Tuch, Hans. *Communicating with the World: U.S. Public Diplomacy Overseas*. New York: St. Martin's Press, 1990.

Tuchman, Barbara W. *The March of Folly*. New York: Ballantine Books, 1984.

U.S. Chairman of the Joint Chiefs of Staff. *Doctrine for Joint Operations (Joint Pub 3-0)*. Washington: Office of the Joint Chiefs of Staff, 1993.

U.S. Department of Defense. *Fiscal Year 2010 Budget Request Summary Justification*, Washington, DC: May 2009.

U.S. Department of Defense. Report of the Defense Science Board Task Force on Strategic Communications, Washington, DC: Office of the Under Secretary of Defense for Acquisition, Technology, and Logistics, September 2004, 2, http://www.acq.osd.mil/dsb/reports/2004-09-Strategic_Communication.pdf.

Vego, Milan N. "Systems versus Classical Approach to Warfare." *Joint Forces Quarterly* 52 (1st Quarter 2009).

Vick, Allen J. *Building Confidence during War and Peace*. Santa Monica, CA: RAND Corporation, 1988.

Volkogonov, Lt. Gen. D. A. *Marksistsko-Leninskoye ucheniye o voyne i armii (Marixst-Leninist teaching on war and the army)*. Moscow: Voyenizdat, 1984.

Wardak, Ghulam Dastagir, comp., and Graham Hall Turbiville, Jr., ed. "The Voroshilov Lectures: Materials from the Soviet General Staff Academy," vol. 1, *Issues of Soviet Military Strategy*. Washington, DC: National Defense University Press, 1989.

Wayne, Jessica. "Operation Just Cause: A Historical Analysis" *Council on Human Affairs*, http://www.coha.org/operation-just-cause-a-historical-analysis/.

Weigley, Russell F. *The American Way of War*. Bloomington, IA: University of Indiana Press, 1977.

Weintraub, Stanley. *Silent Night: The Story of the World War I Christmas Truce*, New York: Plume, 2002.

Wellington, Arthur Wellesley. *The Dispatches of Field Marshall the Duke of Wellington during His Various Campaigns in India, Denmark, Portugal, Spain, the Low Countries, and France from 1799–1818*, Vols 1 and 2 Only. Kessinger Publishing, 2008, at http://www.zum.de/whkmla/period/17891914/x17891914.html.

White House Documents and Publications. "The National Security Strategy of the United States of America." Office of the President of the United States: White House Documents and Publications, March 2006, at http://www.iwar.org.uk/military/resources/nss-2006/index.htm.

White House Press Secretary. "The National Security Strategy of the United States of America." *Whitehouse.gov*. 2006. http://www.whitehouse.gov/nsc/nss/2006/nss2006.pdf, accessed May 6, 2010.

Yager, Harry R. *Strategic Theory for the 21st Century: The Little Book on Big Strategy*. Carlisle: PA, Strategic Studies Institute (monograph), February 2006.

Yates, Lawrence A. *The U.S. Military's Experience in Stability Operations, 1789–2005: Global War on Terrorism Occasional Paper 15,* Fort Leavenworth, KS: Combat Studies Institute Press, 2005.

Yazov, D. T. *Novaya model' bezopasnosti i vooruzhennyye sily (New Model of Security and the Armed Forces). Kommunist* 18 (December 1989).

Zakaria, Fareed. "The Rise of Illiberal Democracy," *Foreign Affairs* (November/ December 1997), http://www.foreignaffairs.com/articles/53577/fareed-zakaria/the-rise-of-illiberal-democracy.

Zartman, William I. *Ripe for Resolution: Conflict and Intervention in Africa.* New York: Oxford University Press, 1990.

Zhurkin, Vitaliy, Sergey Karaganov, and Andrey Kortunov. *Reasonable Sufficiency or How to Break the Vicious Circle. Novoye vremya,* October 2, 1987.

Zhurkin, Vitaliy, Sergey Karaganov, and Andrey Kortunov. *Vyzovy bezopasnosri—staryye i novyye (Challenges to Security: Old and New). Kommunist* 1 (1988): 42–50.

WEB SITES

Brindle at War, http://www.brindle-at-war.net/boerwar.htm, accessed December 25, 2009.

Civil Stabilization Initiative definition, the U.S. Department of State home page, http://www.state.gov/s/crs/66427.html

COIN and FID definitions, U.S. Army home page at http://www.army.mil/-news/2009/11/30/31114-ramping-up-to-face-the-challenge-of-irregular-warfare/.

Counterterrorism definition, the U.S. Department of State home page, http://www.state.gov/s/crs/66427.html.

History.com, this day in history presented by Toyota at http://www.history.com/this-day-in-history/paris-peace-accords-signed.

Imperialism discussion, http://www.uncanny.net/~wsa/iraq3.html.

Korean War Armistice Agreement of July 27, 1953, http://news.findlaw.com/hdocs/docs/korea/kwarmagr072753.html (retrieved December 24, 2009).

Latin American Studies, http://www.latinamericanstudies.org/uruguay/tupa-maros-uruguay.htm.

Machiavelli and Power home page, http://www.emachiavelli.com/Machiavelli%20on%20power.htm.

Millennium Challenge Corporation home page, http://www.mcc.gov/mcc/about/index.shtml.

North African Campaign and a discussion of soldiers' actions during a truce, The Deuce of Clubs Book Club, http://www.deuceofclubs.com/books/177xmas.htm (accessed December 25, 2009, pp. 172–173).

OIF fight termination, CNN.COM/Word, "Bush Calls End to 'Major Combat'" posted May 2, 2003, http://www.cnn.com/2003/WORLD/meast/05/01/sprj.irq.main/.

Power discussion, http://www.newyorker.com/talk/2009/01/26/090126ta_talk_hertzberg.

President George W. Bush's Address Regarding Ultimatum to Iraq, http://www.johnstonsarchive.net/terrorism/bushiraq2.html.

S/CRS information and Organizational Chart found at http://www.crs.state
    .gov/index.cfm?fuseaction=public.display&shortcut=CRPF

Suez Canal information, http://www.jewishvirtuallibrary.org/jsource/History/
    Suez_War.html.

Third World Nationalism, http://www.uncanny.net/~wsa/iraq3.html.

Title 51 discussion, *Washington Post* online at http://www.washingtonpost.com/
    wp-srv/nation/documents/Gates_to_Clinton_121509.pdf.

Tulane University's home page, "Crisis at Fort Sumter" section, http://www
    .tulane.edu/~sumter/Background/BackgroundForts.html.

U.S. Army home page, http://www.army.mil/-news/2009/11/30/31114-ramping-
    up-to-face-the-challenge-of-irregular-warfare/

Index

Abkhazia, 21
Advance Civilian Team (ACT), 63
Al Qaeda, x, 3, 50, 79, 89–90, 123. *See also* AQI, 92, 103–104, 106
Albania, 7, 43–44
Al-Zarqawi, Abu Musab, 90
Afghanistan, x, 4, 8, 15, 18, 20, 22, 49, 51–53, 69, 71, 73, 79–82, 85, 89, 102, 123–128, 130, 132, 135, 143, 165; International Security Assistance Force (ISAF), 52–53; Karzai, Hamid, 125
Alvarez, Gregorio, 102
American Civil War, 31, 34, 39, 102
Anbar Province, 3, 17, 92
Armistice, 13, 26–29, 34–38, 107–114, 117–118, 159; Armistice agreements between Israel and its neighbors Egypt, Jordan, Lebanon, and Syria, 1949, 29
Armistice in Algeria in an attempt to end the Algerian War, 1962, 29
Armistice with Bulgaria, 28
Armistice of Copenhagen of 1547, 28; Armistice with France (Second Compiègne), 1940, 28; Armistice with Germany (signed at Compiègne), 1918, 28; Armistice with Italy, 1943, 29; Armistice of Mudanya between Turkey, Italy, France, Britain, and later Greece, 1922, 28; Armistice of Mudros between the Ottoman Empire and the Allies, 1918, 28; Armistice of Saint Jean d' Acre between British forces in the Middle East and Vichy France forces in Syria, 1941, 28; Armistice of Trung Gia ending the First Indochina War, signed by France and the Viet Minh on July 20, 1954, 29; Austrian-Italian Armistice of Villa Giusti, 28; Japanese Instrument of Surrender on September 2, 1945, 29; Korean War Armistice, July 1953, 29; Moscow Armistice ending the Continuation War, signed by Finland and the Soviet Union on September 19, 1944, 29; Peace of Westphalia of 1648, 28; Treaty of Brest-Litovsk, 28; Unconditional surrender implemented by Germany at the end of the war, immediately prior to V–E Day on May 8, 1945, 29

About the Author

DOMINIC J. CARACCILO has commanded two companies in the 82nd Airborne Division, one in Desert Storm, a 173rd Airborne battalion in combat parachute assault in Northern Iraq, and a 101st Airborne brigade combat team in combat in South Baghdad at the height of the insurgency. He was also the operations officer for the 101st Airborne as they controlled Multinational-Division North in Iraq and a member of the 75th Rangers in Afghanistan during the initial assault in Operation Enduring Freedom in 2001. Most recently, he was the Executive Officer to General Ray Odierno for U.S. Forces-Iraq in Baghdad. He currently serves as the Deputy Commander for the 101st Airborne Division and Acting Senior Commander of Fort Campbell.

Caraccilo received a master's degree in National Security and Strategic Studies from the U.S. Naval Command and Staff College, received a second master's degree in Engineering from Cornell University in Operations Research and Industrial Engineering, and served as an assistant professor at West Point.

He is the author of a number of books and commercial and professional articles including *Achieving Victory in Iraq: Countering an Insurgency* (Stackpole 2008), *Surviving Bataan and Beyond* (Stackpole 1999 and 2005). He was a contributing author to the two-volume set, *The Faces of Victory* (Taylor, 1995) and *The Ready Brigade* (McFarland, 1993).